SANTA CRUZ CITY-COUNTY LIBRARY SYSTEM
0000114412034

DISCARD
D0538270

INTERNATIONAL WILDLIFE ENCYCLOPEDIA

THIRD EDITION

Volume 13

NEW–PAR

Marshall Cavendish Corporation
99 White Plains Road
Tarrytown, New York 10591–9001

Website: www.marshallcavendish.com

Library of Congress Cataloging-in-Publication Data

Burton, Maurice, 1898-
International wildlife encyclopedia / [Maurice Burton, Robert Burton] .-- 3rd ed.
p. cm.
Includes bibliographical references (p.).
Contents: v. 1. Aardvark - barnacle goose -- v. 2. Barn owl - brow-antlered deer -- v. 3. Brown bear - cheetah -- v. 4. Chickaree - crabs -- v. 5. Crab spider - ducks and geese -- v. 6. Dugong - flounder -- v. 7. Flowerpecker - golden mole -- v. 8. Golden oriole - hartebeest -- v. 9. Harvesting ant - jackal -- v. 10. Jackdaw - lemur -- v. 11. Leopard - marten -- v. 12. Martial eagle - needlefish -- v. 13. Newt - paradise fish -- v. 14. Paradoxical frog - poorwill -- v. 15. Porbeagle - rice rat -- v. 16. Rifleman - sea slug -- v. 17. Sea snake - sole -- v. 18. Solenodon - swan -- v. 19. Sweetfish - tree snake -- v. 20. Tree squirrel - water spider -- v. 21. Water vole - zorille -- v. 22. Index volume.
ISBN 0-7614-7266-5 (set) -- ISBN 0-7614-7267-3 (v. 1) -- ISBN 0-7614-7268-1 (v. 2) -- ISBN 0-7614-7269-X (v. 3) -- ISBN 0-7614-7270-3 (v. 4) -- ISBN 0-7614-7271-1 (v. 5) -- ISBN 0-7614-7272-X (v. 6) -- ISBN 0-7614-7273-8 (v. 7) -- ISBN 0-7614-7274-6 (v. 8) -- ISBN 0-7614-7275-4 (v. 9) -- ISBN 0-7614-7276-2 (v. 10) -- ISBN 0-7614-7277-0 (v. 11) -- ISBN 0-7614-7278-9 (v. 12) -- ISBN 0-7614-7279-7 (v. 13) -- ISBN 0-7614-7280-0 (v. 14) -- ISBN 0-7614-7281-9 (v. 15) -- ISBN 0-7614-7282-7 (v. 16) -- ISBN 0-7614-7283-5 (v. 17) -- ISBN 0-7614-7284-3 (v. 18) -- ISBN 0-7614-7285-1 (v. 19) -- ISBN 0-7614-7286-X (v. 20) -- ISBN 0-7614-7287-8 (v. 21) -- ISBN 0-7614-7288-6 (v. 22)
1. Zoology -- Dictionaries. I. Burton, Robert, 1941- . II. Title.

QL9 .B796 2002
590'.3--dc21

2001017458

Printed in Malaysia
Bound in the United States of America

07 06 05 04 03 02 01 8 7 6 5 4 3 2 1

Brown Partworks
Project editor: Ben Hoare
Associate editors: Lesley Campbell-Wright, Rob Dimery, Robert Houston, Jane Lanigan, Sally McFall, Chris Marshall, Paul Thompson, Matthew D. S. Turner
Managing editor: Tim Cooke
Designer: Paul Griffin
Picture researchers: Brenda Clynch, Becky Cox
Illustrators: Ian Lycett, Catherine Ward
Indexer: Kay Ollerenshaw

Marshall Cavendish Corporation
Editorial director: Paul Bernabeo

Authors and Consultants

Dr. Roger Avery, BSc, PhD (University of Bristol)

Rob Cave, BA (University of Plymouth)

Fergus Collins, BA (University of Liverpool)

Dr. Julia J. Day, BSc (University of Bristol), PhD (University of London)

Tom Day, BA, MA (University of Cambridge), MSc (University of Southampton)

Bridget Giles, BA (University of London)

Leon Gray, BSc (University of London)

Tim Harris, BSc (University of Reading)

Richard Hoey, BSc, MPhil (University of Manchester), MSc (University of London)

Dr. Terry J. Holt, BSc, PhD (University of Liverpool)

Dr. Robert D. Houston, BA, MA (University of Oxford), PhD (University of Bristol)

Steve Hurley, BSc (University of London), MRes (University of York)

Tom Jackson, BSc (University of Bristol)

E. Vicky Jenkins, BSc (University of Edinburgh), MSc (University of Aberdeen)

Dr. Jamie McDonald, BSc (University of York), PhD (University of Birmingham)

Dr. Robbie A. McDonald, BSc (University of St. Andrews), PhD (University of Bristol)

Dr. James W. R. Martin, BSc (University of Leeds), PhD (University of Bristol)

Dr. Tabetha Newman, BSc, PhD (University of Bristol)

Dr. J. Pimenta, BSc (University of London), PhD (University of Bristol)

Dr. Kieren Pitts, BSc, MSc (University of Exeter), PhD (University of Bristol)

Dr. Stephen J. Rossiter, BSc (University of Sussex), PhD (University of Bristol)

Dr. Sugoto Roy, PhD (University of Bristol)

Dr. Adrian Seymour, BSc, PhD (University of Bristol)

Dr. Salma H. A. Shalla, BSc, MSc, PhD (Suez Canal University, Egypt)

Dr. S. Stefanni, PhD (University of Bristol)

Steve Swaby, BA (University of Exeter)

Matthew D. S. Turner, BA (University of Loughborough), FZSL (Fellow of the Zoological Society of London)

Alastair Ward, BSc (University of Glasgow), MRes (University of York)

Dr. Michael J. Weedon, BSc, MSc, PhD (University of Bristol)

Alwyne Wheeler, former Head of the Fish Section, Natural History Museum, London

Picture Credits

Heather Angel: 1750, 1782, 1797, 1810, 1842; **Ardea London:** Don Hadden 1867, J.L. Mason 1821, Pat Morris 1749, 1866; **Neil Bowman:** 1775, 1806, 1858; **Bruce Coleman:** Trevor Barrett 1765, 1857, Jen and Des Bartlett 1826, Erwin and Peggy Bauer 1838, Mr. J. Brackenbury 1864, Fred Bruemmer 1814, 1815, Jane Burton 1732, 1733, 1780, 1781, 1800, John Cancalosi 1847, Mr. P. Clement 1734, Bruce Coleman Inc 1816, Bruce Coleman Ltd 1741, Alain Compost 1755, 1855, Sarah Cook 1776, Gerald S. Cubitt 1743, Christer Fredriksson 1799, Bob Glover 1837, Charles and Sandra Hood 1832, 1833, HPH Photography 1766, 1831, Wayne Lankinen 1759, 1761, 1825, Luiz Claudio Marigo 1778, 1819, 1828, Joe McDonald 1793, Rita Meyer 1739, Dr. Scott Nielsen 1851, 1852, Tero Niemi 1758, Pacific Stock 1769, 1772, 1779, William S. Paton 1811, Alan G. Potts 1836, 1737, 1738, Marie Read 1762, 1805, 1807, 1835, Hans Reinhard 1792, 1829, 1869, 1886, John Shaw 1777, Kim Taylor 1760, 1796, 1801, 1827, 1830, 1865, John Visser 1736, 1802, Uwe Walz 1820, Staffan Widstrand 1848, Petr Zabransky 1783, Gunter Ziesler 1747; **Corbis:** Michael and Patricia Fogden 1746, Peter Johnson 1803, Amos Nachoum 1770, 1771, Kennan Ward 1843; **Chris Gomersall:** 1740, 1756, 1812, 1813; **Natural Visions:** Brian Rogers 1808, Slim Sreedharan 1850; **NHPA:** A.N.T. 1824, Daryl Balfour 1849, Bruce Beehler 1823, G.J. Cambridge 1795, Nigel J. Dennis 1862, Robert Erwin 1745, 1844, 1853, K. Ghani 1859, 1860, Martin Harvey 1754, Brian Hawkes 1784, Daniel Heuclin 1735, 1753, 1788, 1804, 1834, Hellio and Van Ingen 1789, E.A. Janes 1742, T. Kitchin and V. Hurst 1757, 1840, 1841, Gerard Lacz 1787, Haroldo Palo 1794, Dr. Ivan Polunin 1863, Jany Sauvanet 1748, Kevin Schafer 1845, Eric Soder 1785, Morten Strange 1856, Norbert Wu 1767, 1773, 1774; **Oxford Scientific Films:** Frithjof Skibbe 1822; **Still Pictures:** Compost/Visage 1798, Michael Gunther 1809, C. Kaiser 1786, Edward Park 1751, 1752, Roland Seitre 1839, Norbert Wu 1846; **Windrush Photos:** 1744.
Artwork: Ian Lycett 1854, Catherine Ward 1763, 1790, 1817, 1861.

Contents

NEWT

A palmate newt sits on a cushion of golden saxifrage. Like all newts, this species breeds in fresh water and is always found in damp places.

NEWTS ARE AMPHIBIANS AND belong to the salamander family. They have a life history similar to that of frogs and toads in that the adults spend most of their life on land but return to water to breed. They are different in form from frogs and toads, however, having long, slender bodies like those of lizards, with a tail that is flattened laterally. The name comes from the Anglo-Saxon *evete*, which became *ewt* and finally *a newt* through the displacement of the *n* in *an ewt*. In Britain, newt refers solely to the genus *Triturus*, but in North America it has been applied to related animals, which are sometimes called salamanders.

Newts of the genus *Triturus* are found in Europe, Asia, North Africa and North America. There are three species native to Britain. The most common is the smooth newt, *T. vulgaris*, which is found all over Europe and is the only newt found in Ireland. The maximum length of the smooth newt is 4 inches (10 cm). The color of the body varies but is mainly olive brown with darker spots on the upper side and streaks on the head. The vermilion or orange underside has round black spots and the throat is yellow or white. The female is generally paler on the underside than is the male and sometimes is unspotted. In the breeding season the male develops a wavy crest running along the back and tail. The palmate newt, *T. helveticus*, is very similar to the smooth newt but about 1 inch (2.5 cm) shorter and with a square-sided body. In the breeding season the males of the two species can be distinguished because black webs link the toes of the hind feet of the palmate newt, and its crest is not wavy. In addition, the tail ends abruptly, and a short thread, about 6 millimeters long, protrudes from the tip. The largest European newt is the crested, or warty, newt, *T. cristatus*. It grows up to 6 inches (15 cm) long. The dark gray skin of the upperparts is covered with warts, while the underparts are yellow or orange and are spotted with black. This species' distinguishing feature, besides its size, is the male's crest, a tall, "toothed" frill that runs from the head to the hips and becomes the tail fin.

Breeding in water

When they come out of hibernation in spring, newts enter the water to breed. They make their way to ponds and other stretches of still water where water plants grow. Newts swim by lash-

NEWTS

CLASS **Amphibia**

ORDER **Caudata**

FAMILY **Salamandridae**

GENUS **4, including *Taricha* and *Triturus***

SPECIES **53, including Californian newt, *Taricha torosa*; crested newt, *Triturus cristatus*; palmate newt, *T. helveticus*; and smooth newt, *T. vulgaris***

LENGTH
Most species less than 5 in. (13 cm)

DISTINCTIVE FEATURES
Long, slender body; long, laterally flattened tail. Breeding male: bright colors; wavy crest along tail and sometimes also down back.

DIET
Various terrestrial and aquatic invertebrates; larger species: also small fish, other small amphibians and amphibian eggs

BREEDING
Breeding season: 2–4 weeks in spring; larval period: 2–4 months

LIFE SPAN
***Taricha torosa*: up to 21 years in captivity; *Triturus vulgaris*: up to 28 years in captivity; much less in wild**

HABITAT
Varies according to species. *Taricha torosa*: mainly grassland, woodland and forest. *Triturus cristatus*: pools, streams and adjacent woods. Breed in ponds (*Taricha* species) or in streams (*Triturus* species).

DISTRIBUTION
North America; most of Europe; western and Central Asia; North Africa

STATUS
Varies according to species; at least one species at low risk, many others common

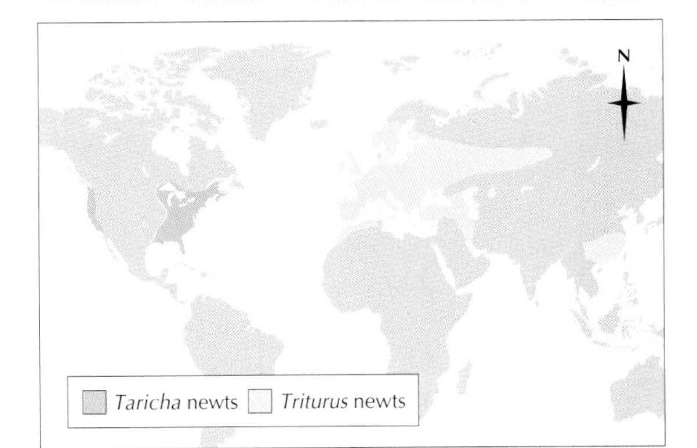

A crested, or warty, newt shows the black-and-orange warning coloration on its underparts. The fact that the orange coloration extends the length of the tail indicates that this individual is a female.

ing with their tails, but they spend much of their time resting on the mud or among the stems of plants. Newts can breathe through their skins, but every now and then they rise to the surface to gulp air. Adult newts do not leave the water immediately after breeding has finished but remain aquatic until July or August. When they return to the land the crest is reabsorbed and the skin becomes rougher. A few individuals remain in the water year-round, retaining their smooth skins and crests. The crested newt keeps its skin moist from the numerous mucous glands scattered over the surface of its body.

Hibernation begins in the autumn, when the newts crawl into crevices in the ground or under logs and stones. They cannot burrow but are very adept at squeezing themselves into cracks. Occasionally several will gather together in one place and hibernate in a tight mass.

Gripping teeth

Newts' jaws are lined with tiny teeth. There are also two rows of teeth on the roof of the mouth. These are not used for cutting food or for chewing but merely to hold slippery, often wriggling prey. Newts feed on a variety of small animals such as worms, snails and insects when on land, and crustaceans, tadpoles and insect larvae while living in water. Unlike frogs and toads, newts do not use their hands to push the food into their mouths but gulp it down with convulsive swallows. Newts swallow snails whole and eat caddis-flies in their cases. Crested newts eat the smaller smooth newts.

Internal fertilization

The mating habits of newts differ from those of common frogs and common toads. Fertilization is internal. The male stimulates the female into breeding condition by nudging her with his snout and lashing the water with his tail. He positions himself in front of or beside her, bends his tail double and vibrates it rapidly, setting up vibrations in the water. Secretions from glands in the male's skin also stimulate the female. At the end of the courtship the male emits a spermatophore, a mass of spermatozoa embedded in a gelatinous substance. The spermatophore sinks to the bottom, and the female newt positions herself over it, then picks it up with her cloaca by pressing her body onto it.

After she has fertilized them, the female lays the 200 to 300 eggs on the leaves of water plants or attaches them to rocks. She usually lays them singly, although some American newts lay their eggs in spherical clusters. In some species, the female newt tests leaves by smell and touch. When she has chosen a suitable one, she holds it with her hind feet and folds the leaf over to form a tube; she then lays an egg in it. The jelly surrounding the egg glues the leaf firmly in place to protect it.

The eggs hatch in about 3 weeks, a tadpole issuing from each one. Newt tadpoles are more streamlined than those of frogs or toads. They are not very different from adult newts except that they have a frill of gills and no legs. Development takes longer than in frog tadpoles, but the young newts are ready to emerge by the end of the summer. A few spend the winter as tadpoles, remaining in the pond until the spring, even surviving being frozen into the ice.

Nasty newts

Newts have many predators: the young are eaten by aquatic insects and the adults by fish, waterbirds, weasels, rats, hedgehogs and many other animals. The crested newt has an unpleasant secretion that is produced in the glands on the back and tail and is exuded when it is squeezed. This secretion deters enemies such as grass snakes, *Natrix natrix*.

Newt's nerve poison

The poison of the crested newt is not only unpleasant, it can also burn the tongue if tasted. However, a far more potent poison is that of the Californian newt, *Taricha torosa*. The poison occurs mainly in the skin, muscles and blood of the newt, as well as in its eggs. Analysis showed that the poison is a substance called tetrodotoxin, which is also found in pufferfish of the family Tetraodontidae. Tetrodotoxin extracted from newts' eggs is so powerful that 0.009 grams can kill 7,000 mice. It acts on the nerves, preventing impulses from being transmitted to the muscles. Somehow, in a manner that biologists do not understand, Californian newts are not affected by their own poison. Their nerves still function when treated with a solution of tetrodotoxin 25,000 times stronger than that which will completely deaden a frog's nerves.

The smooth newt is the most common European newt. In the breeding season the male develops a wavy crest along his back and courts a potential mate in a complex ritual dance.

NIGHT ADDER

THE SIX SPECIES OF night adders belong to the viper family. They are probably more primitive than other vipers, as they lay eggs instead of bearing their young live. The scales on night vipers' heads are comparatively large. Another way in which these snakes differ from other vipers, and from all other venomous snakes, is that the poison glands extend backward into the body. In some species of night adders they may extend more than 4 inches (10 cm) behind the neck and are connected to the fangs by longs ducts. Although night adders are nocturnal, the eyes have round pupils during both night and day.

Night adders live in Africa, mainly south of the Sahara Desert in open country such as savanna. The common night adder, *Causus rhombeatus*, is very abundant in many localities, extending from South Africa to the Sudan and Somalia. The stout body with its fairly short tail is about 2 feet (60 cm) long, but common night adders can grow to 3 feet (90 cm). The scales are smooth and the ground color is light to dark gray or brown. Behind the head there is a dark *V* that points toward the snout, and down the back there is a row of dark, rhomboid-shaped patches from which the snake gets its alternative name of rhombic night adder. The snouted night adder, *C. defilippii*, of southern and eastern Africa, is similar in appearance to the common night adder except that the tip of the snout is turned up and it has a shorter body, usually just 12–15 inches (30–38 cm) in length.

Two other species are restricted to eastern Africa. The velvety, or green, night adder, *C. resimus*, lives in Kenya, Tanzania and Uganda. Like that of the snouted night adder, the snout has an upturned tip. The snake is grass green above and whitish underneath and has a *V* on the head and chevrons running down the back. Lichtenstein's night adder, *C. lichtensteinii*, is confined to Uganda and western Kenya. It is olive green with indistinct markings on the back. Both species grow to a length of 18–20 inches (45–50 cm).

Demon adders

These sluggish snakes are often very abundant on open ground that has been cleared for human settlement. Since they are nocturnal, however, they are not responsible for many serious bites. Even if a night adder does bite a person, the outcome is unlikely to be grave, as the poison is not very strong. Even without treatment the symptoms of swelling, hemorrhage and fever are not usually severe unless the victim is a child or a sick person or unless the snake has been allowed to hang on and inject large quantities of venom. The night adder acquired its nickname "demon adder" not because of its venomous bites but because of its threats: it inflates its body and hisses, putting on a bold, blustering show.

Frogs and toads are unusual foods for vipers, but night adders eat large numbers of them. Some writers have gone so far as to state that night adders eat nothing but frogs and toads, but they do also eat rats and mice. The prey is usually swallowed headfirst, having been held while the venom takes effect.

Adder's love dance

When night adders mate in early spring, the male approaches the female from behind, rubbing his chin and throat over her tail and slowly jerking himself forward so that he moves along her body. The female at first is indifferent and continues to wind forward. Then she begins to move more slowly, throwing her body into loops with the male following every move. After

The velvety, or green, night adder is one of the night adders with large poison glands that extend far back into the body cavity.

A common night adder shows off the dark V *mark behind its head. This species is the most widespread of all the night adders.*

a while the male wraps his tail around the female's body and twists so their cloacas are brought together. After mating, the snakes disengage and go their separate ways.

The females lay a clutch of up to about two dozen eggs, which may take up to 4 months to hatch. In tropical parts of Africa night adders may lay several clutches in one year. All these could result from one mating, as the sperms are stored in the female's reproductive tract. One female night adder kept in solitary confinement produced four fertile clutches in 5 months.

Deadly poison apparatus

Snakes can be classified by their poison apparatus as well as by the more usual method of dividing them into families; snakes with similar poison apparatus may not be at all closely related. Three groups of venomous snakes are named after the form of their fangs. The opisthoglyphs have grooved fangs. *Glyph* is derived from the Greek for "carving" and at one time meant "groove." These snakes are not usually very dangerous as they inject little poison when they bite; the boomslang, *Dispholidus typus*, can be an exception. The second group, the proteroglyphs, includes such dangerous snakes as the cobras, coral snakes and water snakes. They have fangs in which the walls of the grooves are sometimes folded over to form canals through which the venom can flow.

The vipers and pit vipers, which include the rattlesnakes, have fangs that fold back when the mouth is shut. These snakes are the solenoglyphs. Their fangs are always tubular, like hypodermic needles. The night adders belong to this group. They are the only members with a trace of a groove on their fangs, which suggests that the hypodermic fangs of the more typical solenoglyphs derived from fangs with open grooves, the walls of which gradually fused.

COMMON NIGHT ADDER

CLASS **Reptilia**

ORDER **Squamata**

SUBORDER **Serpentes**

FAMILY **Viperidae**

GENUS AND SPECIES ***Causus rhombeatus***

ALTERNATIVE NAME
Rhombic night adder

LENGTH
Up to 3 ft. (90 cm); usually 2 ft. (60 cm)

DISTINCTIVE FEATURES
Stout body with fairly short tail; unlike all other snakes in family Viveperidae has large scales on head (similar to head scales of many nonvenomous snakes); gray or brown overall; dark "V" behind head; row of dark, rhomboid-shaped patches along back

DIET
Mainly frogs and toads

BREEDING
Varies according to location. Breeding season: summer (South Africa); number of eggs: 15 to 26; up to 3 clutches per year.

LIFE SPAN
Not known

HABITAT
Dry, savanna-like habitats

DISTRIBUTION
Mali, east to Somalia, south to South Africa

STATUS
Abundant in many localities

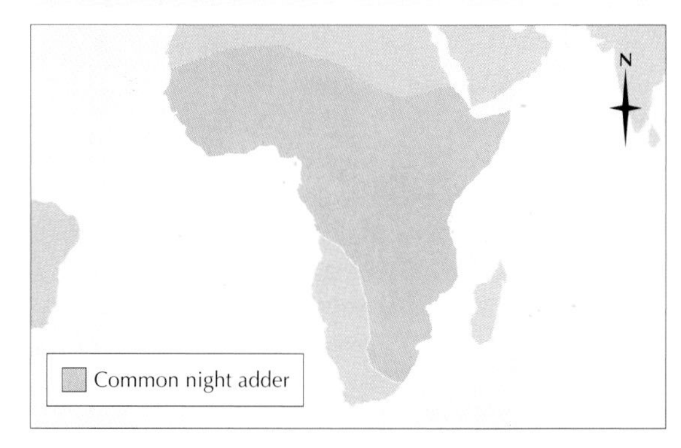

NIGHT HERON

THE NIGHT HERONS ARE a group of medium-sized waterbirds found around the world. They differ from other herons by having shorter legs and necks, and their bills are shorter and heavier. There are eight species, with several more known only as fossils.

The black-crowned night heron, *Nycticorax nyctocorax*, is one of the most cosmopolitan of all birds. It breeds in southern Europe, occasionally wandering northward, as well as in Africa and in central and southern Asia. In America, it breeds from Canada south to Tierra del Fuego and the Falklands Islands. Its plumage is generally gray, darker on the wings and glossy black on the back and crown. It also has prominent red eyes. During the breeding season long white plumes trail from the back of the head. Another species, the yellow-crowned night heron, *N. violacus*, is found throughout South America, and north into California and Texas. In Southeast Asia, Australia and Polynesia, the black-crowned night heron is replaced by the rufous night heron, *N. caledonicus*, also known as the Nankeen night heron. This species is bright chestnut with white underparts. The crown and nape are black and there are three white plumes in the breeding plumage. A white line runs through the eye, under which there is a patch of naked, yellowish-green skin. Many night herons breed in temperate areas, but they migrate to warmer parts of their range over the winter months.

The oriental night herons (genus *Gorsachius*) form a group separate from the other night herons. One species, the white-backed night heron, is widespread throughout sub-Saharan Africa; the others, including the Japanese, Malaysian and white-eared night herons, are found in eastern and southern Asia. These species live mainly in dense forest, and nest solitarily rather than in colonies. *Gorsachius* night herons look quite similar to bitterns.

Active at night

During the day, night herons roost in thick cover, usually on the boughs of leafy trees. When evening falls, the night herons fly off to their feeding grounds, which consist of still or slow-moving fresh water. In parts of the United States the black-crowned night heron is nicknamed the "quawk" or "quok-bird" from its guttural call,

A yellow-crowned night heron searches for fish in the Florida Everglades.

The black-crowned night heron is found over an exceptionally wide range, including a number of isolated island groups. This bird is from West Falkland Island in the southern Atlantic Ocean.

which is often heard when the bird is flying overhead. Its flight is more buoyant than the typical labored beat of larger herons, and the neck is held shortened rather than doubled back. The night herons are strong fliers, and this goes some way towards explaining their exceptionally wide distribution around the globe. Some species have managed to colonize very remote island groups, such as Hawaii, which has a population of black-crowned night herons.

In some of the longer-established island populations, such as the several species of night herons formerly found on Mauritius in the Indian Ocean, the wings evolved to become much reduced, which probably rendered these now-extinct birds flightless.

Fishy diet

Night herons feed mainly on small fish, which they catch by standing motionless and then stabbing with a rapid thrust of the bill. At other times they stalk through the water seeking less active prey. They fish in shallow water, usually about 6 inches (15 cm) deep, near the banks of pools, in reed beds, in bayous and in ditches. Night herons are highly adaptable generalist feeders: as well as fish they also eat frogs and other amphibians, insects, small crustaceans, snails and worms. Even small snakes and mice may occasionally be caught, as well as nestling birds.

BLACK-CROWNED NIGHT HERON

CLASS **Aves**

ORDER **Ciconiiformes**

FAMILY **Ardeidae**

GENUS AND SPECIES ***Nycticorax nycticorax***

WEIGHT
1–1¾ lb. (500–800 g)

LENGTH
Head to tail: 22¾–25½ in. (58–65 cm); wingspan: 41⅓–44 in. (1.05–1.12 m)

DISTINCTIVE FEATURES
Medium-sized heron with stocky body and stout bill. Adult (all year): white underparts; gray upperparts except for black back; black crown and bill; red eye; orange-yellow legs. Adult (breeding): long white plumes trail back from crown; red tinge to legs.

DIET
Amphibians, fish, shrimps, crabs, insects and worms; occasionally small snakes and mice

BREEDING
Age at first breeding: 2–3 years; breeding season: varies across wide range; number of eggs: 3 to 5; incubation period: 21–22 days; fledging period: 40–50 days; breeding interval: 1 year

LIFE SPAN
Up to 20 years

HABITAT
Margins of still or slowly moving fresh water; nests and roosts in trees

DISTRIBUTION
Widespread in North and South America; from southern Europe, east through central and southern Asia to Japan and Indonesia; throughout Africa; also on oceanic islands

STATUS
Common

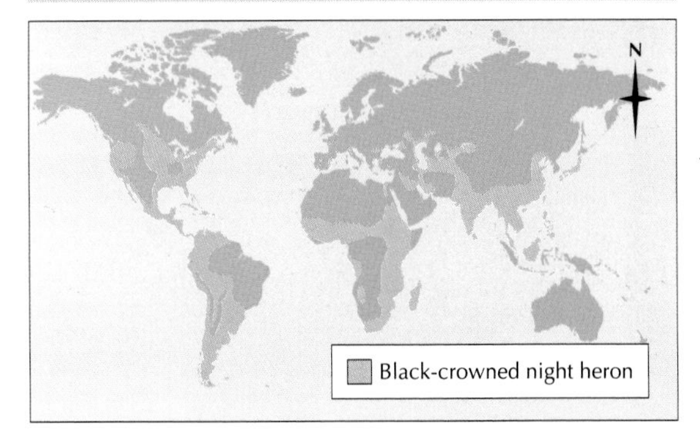

Noisy treetop nests

Apart from the oriental night herons, which nest solitarily, night herons nest in large colonies, with as many as 2,000 nests. There are often as many as 30 nests in each tree, and a single colony may cover many acres of woodland or thicket. At the height of the breeding season, when the colony is packed with adults and young, the noise has been described as being like "several hundred people choking each other." The sight and sound of the colony is made more impressive because night herons often nest alongside egrets, ibises, cormorants and other herons.

It pays to advertise

In the breeding season the legs of the black-crowned night heron, which are usually orange yellow, become tinged with red, and the plumes on the crown become longer. The male chooses the nest site and starts to build the nest, at the same time advertising for a mate.

The male attracts the female by his courtship dance, in which he lowers his head and wings and "marks time" on the branch. At intervals he lowers his head and utters a hissing call. When a female approaches, his head and neck feathers are raised up, the three plumes are spread out so they are almost at right angles to each other and the eyeballs are protruded to show off the red irises. The male then lowers his bill and stretches out his neck toward the watching bird. This greeting ceremony is performed throughout the breeding season whenever the two partners meet, and they also caress each other delicately with their bills.

The nest is a crude platform of twigs and reeds that is sometimes blown down in a high wind. It is usually built in trees or saplings that are growing in or hanging over water. The rufous night heron sometimes nests on the ground, arranging only sufficient sticks to prevent the eggs from rolling around.

Three to five bluish green eggs are laid and incubated by both parents for 3 weeks. Newly hatched chicks are fed on liquid, digested food and after 3 weeks, on semidigested fish and crustaceans. Later the young can seize whole fish from the parents' bill and eventually the parents lay food on the nest for the chicks to pick up. During this time, night herons are more active during daylight hours as they have to find extra food for the chicks.

When they are between 2 and 3 weeks old, the chicks leave the nest and scramble around among the branches, but they do not begin to fly until they reach 6–7 weeks old. Over the next 3 years they gradually attain adult plumage, before beginning to breed themselves.

A black-crowned night heron waits patiently on a log for prey to come into range. These birds seize a variety of prey, including fish, frogs, newts and insects.

NIGHTINGALE

The male nightingale's characteristic song, an advertisement to females, is often heard after dusk.

THE NIGHTINGALE IS A very shy bird that seldom emerges from the undergrowth when feeding. Consequently, its tuneful song is more familiar to most people than the sight of the bird itself. Male and female nightingales are alike in appearance, both about 6½ inches (16.5 cm) long, with russet brown upperparts, dull white tinged with brown below, and with a bright rufous tail and rump. The dull plumage enables the nightingale to blend in effectively with the leaf litter and shrubs of its natural habitat.

The nightingale's summer range extends from England (south and east of a line joining The Wash to the east and the Severn River to the west) eastward across western Europe to the Balkans, Asia Minor and Central Asia, and southward to northwestern Africa. On the outskirts of this range it is a local and irregular visitor. It arrives in Europe in spring and leaves at the end of August and during September, wintering in tropical Africa. There are two allied subspecies in Asia that winter in East Africa.

Terrestrial feeder

The nightingale frequents thick woodland undergrowth, thickets and scrubland and also damp, marshy spots where insects are plentiful. It feeds mainly on the ground, on small invertebrates such as worms, spiders and insects, especially beetles, and also on the larvae of butterflies and moths and the pupae of ants. The nightingale also takes fruit, including berries.

Speckled chicks

The male nightingale usually begins to display to the female in mid-April. He spreads out his rufous tail, rapidly moving it up and down, and flutters his wings with his head dipped. The bulky cup-shaped nest, built by the hen alone, is made chiefly of dead leaves, especially oak leaves, and is lined with dead grass and some feathers. It is built on or a little above the ground in woods or thickets among the brambles and nettles or in hedges. The eggs, usually four or five in number, are olive green or olive brown and are incubated by the hen alone for 13–14 days. During this time the male forages in the surrounding undergrowth for food, which he brings to the female. After hatching, the young are fed by both male and female. They fledge after 11–12 days but are fed by the parents for some days after this. The young have a mottled, speckled appearance rather like that of young European robins but with rufous tails. There is only one brood a year. When the nestlings are fully fledged, the parents often divide the brood, each taking responsibility for the rearing of half the young.

Known for its song

The nightingale is familiar to most people for the power of delivery, richness and variety of its song. The song is characterized by an intermingling of hard and soft notes followed by a single piping note, beginning softly and rising to a crescendo. The volume of the song is surprising considering the size of the bird itself. Sometimes, a sharp, staccato croak occurs in the middle of the melody, which also serves as the nightingale's alarm call.

The male nightingale sings mainly from mid-April to mid-June. It sings during the day as well as at night, although its song is less distinguishable among surrounding birdsong during the daylight hours. During the day, the male sings from several perches, regularly changing location with the start of the next song. By contrast, its nocturnal song is mainly given from one perch. The nightingale also occasionally sings in flight.

NIGHTINGALE

CLASS **Aves**

ORDER **Passeriformes**

FAMILY **Turdidae**

GENUS AND SPECIES ***Luscinia megarhynchos***

WEIGHT
6/10–8/10 oz. (17–24 g)

LENGTH
Head to tail: about 6½ in. (16.5 cm); wingspan: 8–10¼ in. (23–26 cm)

DISTINCTIVE FEATURES
Adult: thin bill; unmarked plumage; rich rufous brown upperparts; white or buffish underparts; reddish rump and tail. Juvenile: speckled brown and white.

DIET
Mainly small ground-living invertebrates, sometimes also flying insects; fruits and berries, especially in fall

BREEDING
Age at first breeding: 1 year; breeding season: eggs laid late April–May; number of eggs: 4 or 5; incubation period: 13–14 days; fledging period: 11–12 days; breeding interval: 1 year

LIFE SPAN
Up to 8 years

HABITAT
Thickets and woodland with dense undergrowth, often near water; also marshy scrub and dry, scrubby hillsides

DISTRIBUTION
Breeding: northwestern Africa; much of western and southern Europe; parts of Turkey, Iran and Central Asia. Winter: sub-Saharan Africa.

STATUS
Common

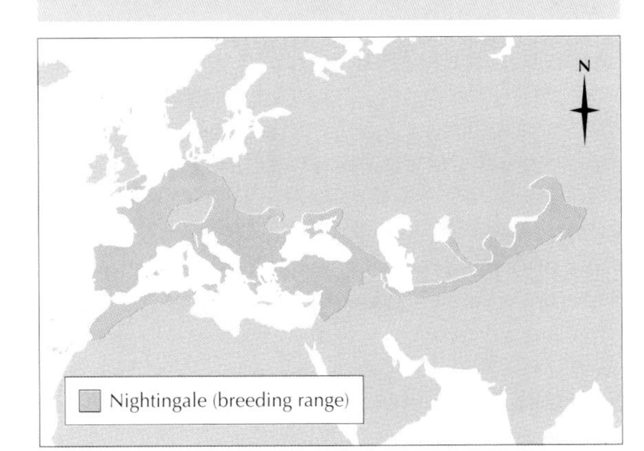

Nestling nightingales grow quickly and are fully fledged in 11–12 days. During this time, the parents take turns feeding them.

Nocturnal singer

Many species of birds sing at night during their breeding season, especially in locations where there are powerful lights, such as along a highway. Their instinct to sing is triggered by the brightness of the lights as it would be by sunlight. A roosting bird will also sometimes sing if it is disturbed, but few birds sing as consistently at night as the nightingale.

One reason for the nightingale's song is to define its territory. Nightingales have small territories, and the male must sing frequently and loudly in order to establish his claim to a certain area. If he is not heard, the male faces competition for food and nesting sites from other birds. However, there is no evidence that nightingales maraud or try to dispossess each other of a territory at night, any more than other songbirds do.

One purpose of the male's song may be to draw the attention of a potential predator away from the hen and nest. However, this seems very unlikely as nesting does not usually begin until mid-May, a full month after the singing has usually begun. Moreover, the song begins to die down in early June, soon after the young are hatched, and at the very time that any decoy value it might have would be of greatest value.

Ornithologists believe that the purpose of the diurnal (daytime) song is mainly to define territory. By moving from perch to perch, the nightingale is able to do this more effectively and is also more likely to avoid confrontation with rivals. The nocturnal song is primarily an advertisement to females. Nightingales fly at night during their annual migration northward from Africa to southern Europe and Asia, and it is probable that the male sings under cover of darkness to attract the attention of migrating females.

NIGHTJAR

An excellently camouflaged Eurasian nightjar (center of photograph) and its two chicks sit on their nest among the leaf litter and bracken.

NIGHTJARS ARE MORE OFTEN heard than seen. They start to fly soon after sundown and are active throughout the night. By day their remarkable camouflage keeps them hidden. Yet despite this, nightjars have probably been given more common names than any other bird. Fern owl, churn owl, eve-jarr, dorhawk and nighthawk are only a few of them. One traditional name is goatsucker.

The best-known nightjar, the Eurasian nightjar, was called *caprimulgus* (goatsucker) by the Romans, and this name is perpetuated today in the bird's scientific appellation, *Caprimulgus europaeus*. The Eurasian nightjar winters in Africa as far south as the Cape and spends its summers in Europe and Asia. It is 10¼–11 inches (26–28 cm) long, and its plumage is gray, barred with buff, chestnut and black. Its bill is small, although the gape is very wide, with strong bristles around the mouth. Adult males have white patches on the tail and wings.

There are 89 species of nightjars and nighthawks, the latter being found only in North and South America. The genus *Caprimulgus* has 56 species. All are superficially similar, including the American poorwill, whippoorwill, chuck-will's-widow and nighthawks, which are dealt with elsewhere. In southern Europe and western North Africa is the red-necked nightjar, *C. ruficollis*, which has a distinctive reddish collar. The Egyptian nightjar, *C. aegyptius*, of southwestern Asia and northern Africa, lives in deserts and is sandy colored. In eastern and Southeast Asia are the jungle nightjar, *C. indicus*, and the large-tailed nightjar, *C. macrurus*. Africa's long-tailed nightjar, *Scotornis climacurus*, has very long central feathers in the tail. The pennant-winged nightjar, *Macrodipteryx vexillarius*, 11 inches (28 cm) long, has the innermost pair of primaries 2 feet (60 cm) long, while the male standard-winged nightjar, *M. longipennis*, in the breeding season has one bare-shafted feather in each wing 2 feet (60 cm) long and a 6-inch (15-cm) flaglike vane at the tip. In Australia the white-throated nightjar (*Eurostopodus mystacalis*), spotted nightjar (*E. guttatus*) and large-tailed nightjar (the latter of which also appears in Asia) are similar to the Eurasian nightjar but larger.

EURASIAN NIGHTJAR

CLASS **Aves**

ORDER **Caprimulgiformes**

FAMILY **Caprimulgidae**

GENUS AND SPECIES ***Caprimulgus europaeus***

ALTERNATIVE NAMES
Many, including fern owl, churn owl, eve-jarr, dorhawk and goatsucker; related American species are called nighthawks

WEIGHT
2⅓–3½ oz. (65–100 g)

LENGTH
Head to tail: 10¼–11 in. (26–28 cm); wingspan: 22½–25¼ in. (57–64 cm)

DISTINCTIVE FEATURES
Very small bill, but with huge gape; long wings and tail; mainly brown plumage, with variety of subtle markings; small white flashes on wings and tail (adult male only)

DIET
Mainly moths and beetles

BREEDING
Age at first breeding: 1 year; breeding season: eggs laid late May–June; number of eggs: 2; incubation period: 17–18 days; fledging period: 16–17 days; breeding interval: 1 or 2 broods per year

LIFE SPAN
Up to 8 years

HABITAT
Dry areas such as open woodland, heather moors, steppe, semidesert and lowland heath

DISTRIBUTION
Breeding: most of Europe, east across Middle East and Asia to northern China. Winter: sub-Saharan Africa.

STATUS
Locally common, but declining in Europe

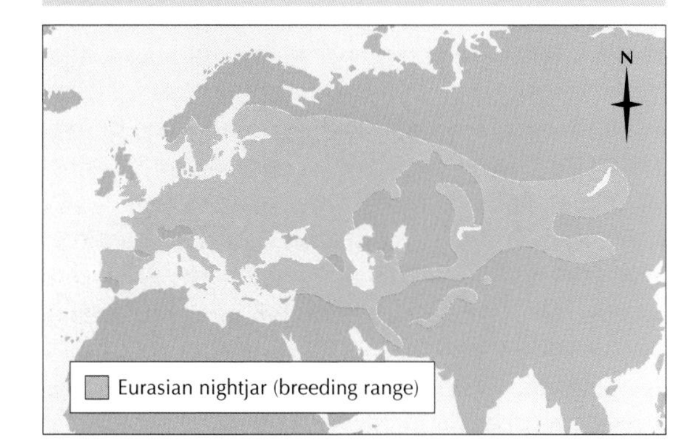

The large-tailed nightjar lives in Asia and Australia. The male's song consists of hollow chopping sounds, rather like an axe being struck against a log.

Cloak of invisibility

The Eurasian nightjar is almost invisible during the day as it rests on the ground on heaths, on bracken-covered slopes and in open woodlands. To say that its plumage harmonizes with its background is less correct than to say that its colors are so broken up that the bird defeats the eye and appears to dissolve into nothing. Whether a particular species of nightjar is among bracken, on lichens, on rocks or on sand, it is almost impossible to see, and it is possible almost to tread on it before it moves. The invisibility is increased by the nightjar closing its eyes and watching an intruder through slits, so its large eyes do not give away its presence. By contrast, the rays of a flashlight shone into bushes at night may be reflected in a pair of red-glowing eyes suggestive of a large animal.

Insect-catcher number one

Nightjars become active at about sundown. They fly with a silent, almost mothlike flight, with strong, deliberate wingbeats alternating with graceful glides and easy, wheeling movements. The characteristic churring or jarring call of a Eurasian nightjar is heard most when the sky is

A common nighthawk perches on a fence post in Colorado. Although they are usually nocturnal, common nighthawks are sometimes also active during the day.

clear. When the skies are overcast, the birds are silent. Nightjars call, almost invariably, while perched lengthwise along a branch.

It used to be said that a nightjar flies with mouth agape, the bristles around the mouth acting as a sweep net to direct insects into it. This belief is now disputed. The nightjar's food is almost wholly insects, from moths and large beetles to mosquitoes. One nightjar examined had 500 mosquitoes in its stomach. Whether the bristles also act as organs of touch, as has been suggested, is problematical. On the undersurface of the nightjar's third toe is a series of sawlike notches. These are used to comb the bristles around the mouth and, so it is said, to remove the scales of moths that have been caught.

Wing claps like pistol shots

The white patches on the wings and tail of the male Eurasian nightjar stand out in courtship. This probably helps the recognition of the sexes. The male flies around the female in wide circles, either beating his wings or holding them stiffly and obliquely over his back, with his tail depressed and fanned. Every now and then he claps his wings with a sound like a pistol shot. Males of other nightjar species use their long feathers for display. The standard-winged nightjar, for example, holds his two flagged primaries vertically over his back, whereas in normal flight they are trailing.

At the end of May the female Eurasian nightjar lays two elliptical eggs, creamy white mottled with brown and purple. No nest is made. The eggs lie on the ground in a shallow, unlined scrape, where both male and female incubate them for 17–18 days; the male takes his turn at dusk and dawn. The parents feed the young birds for a further 16–17 days.

When disturbed at the nest, the parent nightjar performs a distraction display, flopping over the ground as if with a broken wing. When this display is performed in the still of night, the beating of the wings on the ground sounds uncannily loud. After the eggs are hatched, the distraction display is still used to draw the attention of an intruder away from the young birds, which stay very still. Should the predator persist and go near the young, however, they spread their wings, open wide their mouths and lunge at it: a disconcerting bluff.

Egg-carrying parents

The celebrated American ornithologist John J. Audubon described seeing a chuck-will's-widow, *Caprimulgus carolinensis*, an American nightjar, remove its eggs with its mouth when danger threatened. He saw one of the pair wait beside two eggs that had been disturbed, then he saw the two parents each take an egg in the mouth and fly away with it. Audubon's story was long doubted, but this same behavior has been seen since in other species of nightjars.

Legend and the nightjar

Few birds show themselves in as many varying ways as nightjars. One may fly past your face in semidarkness in smooth, silent, almost ghostlike flight. A pair may circle over your head in courtship flight, when the crack of the male's wings has a startling quality in the gathering gloom. In semidesert areas it is not unusual to see a nightjar lying on the ground in the headlights of a car. Usually, even on the brightest moonlit night, it is impossible to see the bird or to track it as it moves from one perch to another, churring first here, then there, like a restless, unseen spirit. It is not surprising, therefore, that some of the names given to nightjars show them to have been held in superstitious awe. The one name that has persisted over the past 2,000 years at least, in numerous languages and in many parts of the world, is goatsucker. The legend is that the nightjar takes milk from the udders of goats. This may be a fanciful allusion to the bird's wide gape. It may even have been a name that sprang from an entirely different source, now obscured by the mists of time.

NIGHT LIZARD

NIGHT LIZARDS LOOK RATHER like geckos and it used to be thought that they were closely related. Also in common with geckos is the permanent "spectacle," composed of a transparent scale, which covers each eye; a soft skin with very small scales on the back; and large eyes that have a vertical pupil by day. However, unlike geckos, night lizards have no pads on the toes to enable them to cling to vertical surfaces, and they produce living young. The 16 species belong to a family called the Xantusiidae, and are found from the southwest of Texas to Panama, with a single species occurring in Cuba.

The desert night lizard, *Xantusia vigilis*, is one of the commonest lizards in the southwestern United States. It is found in the drier parts of California, in southern Nevada and in adjacent parts of Utah and Arizona. It also extends into Baja California and into the deserts of northern Mexico. Despite its name, the desert night lizard is not always nocturnal. It may also be active by day, especially during spring and the fall. It is a small lizard, never more than 3½ inches (9 cm) in total length and may be olive, gray or dark brown in color, with a speckling of black.

Living beneath Joshua trees

The desert night lizard is most often found under dead vegetation at the base of Joshua trees and other cacti, yuccas and agaves, although it also occurs beneath rocks and boulders and under the bark of digger pines in parts of California. A favorite hiding place of these lizards is in the nests that pack, or wood, rats build at the bases of the Joshua trees. The nests consist of piles of sticks and leaves, and so provide protection from predators and good cover from the sun's rays. The lizards may make their homes in the nests while the pack rats are still in residence.

Rocky habitats

Other species of night lizards live in rock crevices or among boulders. For example, the Arizona night lizard, *X. arizonae*, resembles the desert night lizard, but grows to a larger size and is found mostly in rocky places, where it is adept at creeping into small spaces beneath boulders. The granite night lizard, *X. henshawi*, which is confined to a small part of southern California and Baja California, has a pale-colored body with large, dark spots on the back. It is most often found under slabs that have flaked off the bedrock. The Cuban night lizard, *Cricosaura typica*, lives among loose limestone boulders.

The island night lizard, *Klauberina riversiana*, is the largest of the species found in the United States, reaching up to 8 inches (20 cm) in total length. It also has the most restricted distribution and is confined to just three small islands off the coast of southern California. Although largely nocturnal, island night lizards are often seen to be active by day.

Most night lizards, such as this granite night lizard, have highly specialized habitats. Often a single species is confined to a fairly small, local area.

In the summer night lizards lie under logs or leaves, usually in single pairs, although in winter they may gather in groups of up to 40.

Some species of night lizards are quite abundant in the specialized habitats where they occur. However, desert night lizards depend on the survival of Joshua trees, and island night lizards are vulnerable because the islands where they occur are small in size.

Feeding in leaf litter

Joshua trees also supply the desert night lizard with a plentiful source of food because night lizards feed on small animals that live in leaf litter and under logs. Ants, beetles and flies are their favorite prey but they also take moths, spiders and sometimes scorpions. The exception is the island night lizard, which relies heavily on plant material, especially seeds, in order to survive in its relatively barren habitat.

Produce live young

Night lizards are viviparous, meaning that their young are born alive. After fertilization, the eggs take 90–120 days to develop. During this time food passes to the developing embryos and excretory products pass back to the mother through a placenta formed from the joining of the embryonic membranes and the oviduct. In desert night lizards the litter is very small, consisting usually of two or three young. They are born in September or October and measure just under 1 inch (2.5 cm) in total length. Most other species have larger litters than this. For example, in the island night lizard the typical litter size is four to nine young.

In spite of the name, night lizards are not always nocturnal and may be seen during the day. The 16 species are unusual among lizards in that they produce live young.

NIGHT LIZARDS

CLASS **Reptilia**

ORDER **Squamata**

SUBORDER **Sauria**

FAMILY **Xantusiidae**

GENUS ***Xantusia; Klauberina; Gaigeia; Lepidophyma; Cricosaura***

SPECIES **Desert night lizard, *Xantusia vigilis*; Arizona night lizard, *X. arizonae*; island night lizard, *Klauberina riversiana*; others**

LENGTH
Desert night lizard: up to 3½ in. (9 cm); island night lizard: up to 8 in. (20 cm)

DISTINCTIVE FEATURES
Small, mostly nocturnal lizards; large eyes with vertical pupil and no eyelids; velvety skin with very small dorsal scales; resemble geckos but lack adhesive pads on toes

DIET
Invertebrates such as ants, beetles and flies; island night lizard: mainly plant material

BREEDING
All species are viviparous (produce live young). Number of young: 2 or 3 (desert night lizard), 4 to 9 (island night lizard); hatching period: 90–120 days.

LIFE SPAN
Not known

HABITAT
Desert night lizard: at the base of Joshua trees and other cacti, yuccas and agaves; other species: rocky places, among boulders

DISTRIBUTION
Southwestern U.S. south to Panama; single species in Cuba

STATUS
Some species abundant; island night lizard: vulnerable

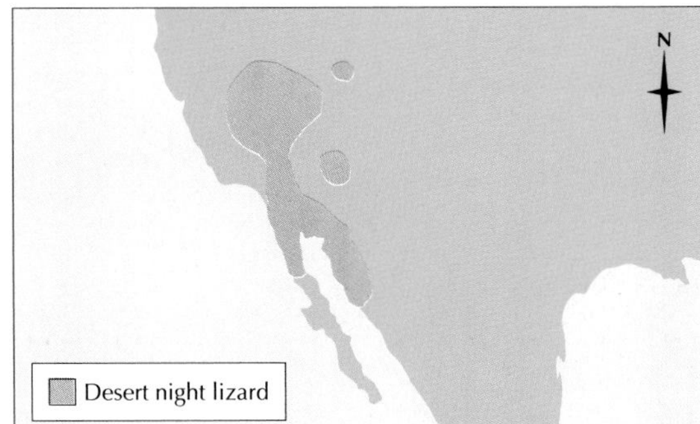

NIGHT MONKEY

Also called the douroucouli, the night monkey is the only nocturnal New World monkey. Its head and body measure 9½–14½ inches (24–37 cm), with a slightly longer tail. The fur is short, soft and woolly, silvery or dark gray above and gray or brown beneath. The head is round, with the short ears almost hidden in the fur, and the face is flat and light gray in color, marked with variable black or brown lines. The retina of the night monkey's eye is made up of rod cells only, a common feature in nocturnal animals, enabling the species to see farther in dim light.

Their long, slender fingers and opposable thumbs make night monkeys highly dextrous. They manipulate their food carefully, holding it in front of their eyes for careful inspection. Wings and legs of insects are bitten off before the head is removed and the body is eaten at leisure.

Nocturnal noise-makers

Night monkeys make loud and resonant calls that echo around forests at night. The calls are produced with the aid of an inflatable pouch under the chin, which is connected to the windpipe. They have a repertoire of some 50 calls, variously described as sounding like owls' hoots, the roars of jaguars, the mewing of cats or the barking of mastiff dogs, as well as various squeaks and chitterings.

Because they are nocturnal animals, night monkeys have not been studied as much as other monkey species. They live in the tall Central and South American rain forests, from Nicaragua to northeastern Argentina, and from Brazil in the east to Peru and Ecuador in the west.

During the day a night monkey sleeps with its head resting between its arms, and the arms and legs are tucked under its body. Families sleep together in holes in tree trunks. If they remain uninterrupted, night monkeys sleep until twilight, at which time they come out to feed.

Silent and agile

The night monkey is able to move through the trees in absolute silence. It runs along branches on all fours and leaps from one branch to another, judging the distance and landing place accurately in what would be pitch darkness to humans. Accurate jumping is assisted by the long tail, which streams out behind the monkey and acts as a balance or rudder, in the manner of

Night monkeys are sometimes called owl monkeys because they utter owl-like hoots and have striking markings around their large eyes.

The night monkey's long fingers and thumbs enable it to grip tightly onto branches and trunks.

a squirrel's tail. Under the tail the hair is long and stiff, forming a keel that increases its efficiency as a rudder. The tail is also used as a brake. As the night monkey lands, the tail is swept forward beneath its body and under the branch, the tail's stiffness slowing down and steadying the night monkey's descent.

Their good night vision enables night monkeys to catch even moving animals in the dark. They can catch flies and moths as they fly past and also take spiders, small birds and small mammals, as well as fruits, nuts and leaves.

Breeding in captivity

Little is known about the breeding habits of night monkeys in the wild. They are usually seen sleeping or feeding in small family parties, made up of the two parents with one or two young. Occasionally the parties are larger, and zoologists believe that this happens when young night monkeys that have grown up remain with their parents, helping to take care of younger offspring. The evidence for this theory is that the pattern of fur color on night monkeys varies considerably from one animal to another, but in these large family parties the patterns are similar, suggesting that the individuals may be related.

There are some records of night monkeys breeding in captivity. In one sample case a baby was born weighing 3½ ounces (100 g). It was helpless, being unable to do more than cling to its mother, but grew rapidly, putting on 1 ounce (28 g) each week, and at 4 weeks could leave its mother and walk for short distances. The youngster continued to spend much of the time on its mother until 7 weeks old. At this age it started leaping from branch to branch and by the age of 2 months it was almost fully grown.

NIGHT MONKEYS

CLASS **Mammalia**

ORDER **Primates**

FAMILY **Cebidae**

GENUS ***Aotus***

SPECIES ***Aotus lemurinus; A. brumbacki; A. hershkovitzi; A. trivirgatus; A. vociferans; A. miconax; A. nigriceps; A. nancymaae; A. infulatus; A. azarai***

ALTERNATIVE NAMES
Douroucouli; owl monkey

WEIGHT
1⅓–2¼ lb. (0.6–1 kg)

LENGTH
Head and body: 9½–14½ in. (24–37 cm); tail: 12¼–15¾ in. (31–40 cm)

DISTINCTIVE FEATURES
Soft, dense fur; dark gray above, lighter below; darker lines either side and between eyes; white markings above eyes; small, round head; very large eyes

DIET
Nuts, fruits, leaves, bark, flowers, insects, spiders and small vertebrates

BREEDING
Age at first breeding: 3 years; breeding season: all year, mainly November–January; number of young: usually 1; gestation period: about 135 days; breeding interval: about 13 months

LIFE SPAN
Up to 27 years in captivity

HABITAT
Forests

DISTRIBUTION
Nicaragua south to Argentina

STATUS
Varies according to species; none endangered

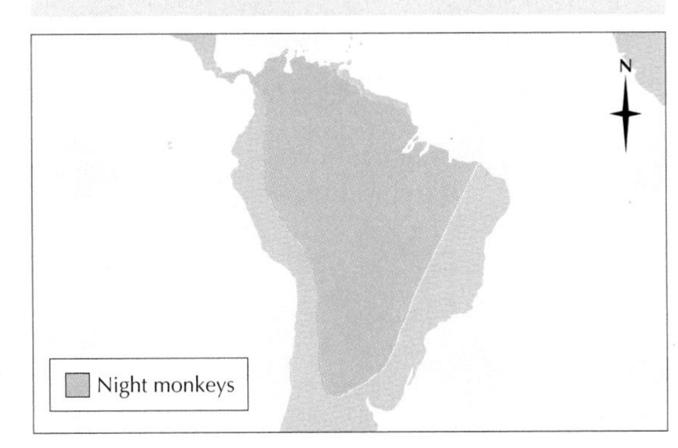

NILE FISH

Several fish living in the Nile and in rivers in tropical West Africa have the ability to swim backward and forward with equal ease. In the mid- to late 20th century, scientists also found that the fish generate their own electrical field. One of these fish is *Gymnarchus niloticus*, commonly known as the Nile fish.

The Nile fish can reach up to 5 feet (1.5 m) in length. Its body is flattened from side to side and ends in a slender "rat's tail." The only fins it has are a pair of very small pectorals and a long ribbonlike dorsal fin starting just behind the head and ending well short of the slender tip of the body. The head is rounded, with small eyes, a blunt snout, a wide mouth and strong teeth. The body is covered with very small scales.

Some authorities class the Nile fish as the only species in the family Gymnarchidae. Others place it in the family Mormyridae. The latter consists of 18 genera divided into 198 species and includes the elephant-nose fish, *Gnathonemus petersii*. These species have a more conventional fish shape. The body is compressed, and although the pelvic fins are absent and the pectoral fins are small, both dorsal and anal fins are well developed and the anal fin is usually longer than the dorsal fin. The tail fin is forked. Some species have a fingerlike process (projection) on the jaw that is used as a feeler for searching out small animal food in the mud. Others have a long proboscis-like chin that gives the fish a striking appearance. Mormyrids swim by waving the dorsal and anal fins and keeping the body rigid. Like the Nile fish, they also have a weak electrical system, with which they sense objects around them.

The Nile fish ranges from the upper reaches of the River Nile to the Chad Basin and beyond to Senegal and the basin of the Niger River. Mormyrid fish occupy much the same area but also spread into the rivers of the Democratic Republic of the Congo (Zaire).

Plugging into a fish

The Nile fish possesses an electrical organ that extends along almost the entire length of its body, to the tip of its tail. The discovery of an electrical field surrounding the Nile fish was first published in 1951. Muscles on either side of the tail form electrical generators that constantly emit pulses at the rate of 300 a second. A scientist found that when he lowered a pair of electrodes connected to an oscilloscope into water containing a Nile fish, the oscilloscope registered the fish's electrical discharges.

Each discharge spreads out through the surrounding water, forming an electrical field similar to the field around a bar magnet, with the positive pole at the fish's head and the negative pole at its tail. Any object in the water disturbs the field, as scientists found when they experimented by dipping the two ends of a U-shaped copper wire, a material that conducts electricity, into the field near the fish. The fish was disturbed and swam away, but would remain still and undisturbed if a similarly shaped, nonconducting material was lowered into the water.

Surrounded by an electrical field

The sense organs with which Nile fish pick up disturbances in their electrical field are minute jelly-filled pits in the skin of the head, each with a receptor at the bottom. The receptors may be pouchlike or tuberous in form and are similar to the lateral line sense organs of other fish.

In experiments to test the fish's receptivity to electrical impulses, scientists placed two porous clay tubes into an aquarium containing Nile fish. One tube was filled with tap water or a similar conductor and the other with a nonconductor such as wax or glass. The fish was trained to come to the conducting tube and soon learned to ignore the nonconductor. By changing the

The Nile fish generates electricity through organs in the rear of its body, sensing disturbances in the electrical field through receptors in the head region.

The elephant-nose fish is distinguished by its forked tail fin and, despite its name, an elongated jaw. Like the Nile fish, it has an electrical organ and can swim backward.

contents of the clay tubes, the Nile fish's sensitivity was revealed. It was able to detect the presence in a tube of a glass rod 2 millimeters in diameter, which caused only a minute change in the fish's electrical field. When the Nile fish's own electrical discharges were recorded and played back to it, the fish attacked the electrodes as if they were another fish.

A fish that cannot bend

The electrical field enables the Nile fish to detect the small fish on which it feeds without having to see them. Visibility is poor in the muddy and often dark waters that the Nile fish inhabits, and it does not generally use its eyes to sense objects. The fish can confidently move backward as well as forward because any obstacles in its path disturb the pattern of its electrical field.

The Nile fish swims in a smooth glide with a rigid body, driven by a wave passing through the long dorsal fin. To swim backward the fish merely reverses the direction of this wave. It is necessary for the fish to preserve a rigid body when it swims, because if it were to move its body, as most fish do when swimming, the electrical field around it would be disturbed.

Floating home

The Nile fish makes a floating four-sided nest of grass and other vegetation, with three sides out of the water and the fourth submerged to a depth of 4–8 inches (10–20 cm). The female lays about 1,000 large, amber-colored eggs. The larvae have long gill filaments and retain the remains of the yolk sac. They stay in the shelter of the nest for 5 days, by which time the yolk is used up and the young fish are about 3 inches (7.5 cm) long.

NILE FISH

CLASS **Osteichthyes**

ORDER **Osteoglossiformes**

FAMILY **Gymnarchidae**

GENUS AND SPECIES ***Gymnarchus niloticus***

ALTERNATIVE NAME
Aba

WEIGHT
Up to 42 lb. (19 kg), usually much less

LENGTH
Up to 5 ft. (1.5 m)

DISTINCTIVE FEATURES
Long body, flattened from side to side; almost cylindrical head with blunt snout; wide mouth; short, strong teeth in single row in both jaws; very small eyes; long, ribbonlike dorsal fin; short, rounded pectoral fins; blackish, or steel gray in color; electrical organ extends along most of body

DIET
Crustaceans, insects and small fish

BREEDING
Number of eggs: about 1,000; hatching period: 5 days

LIFE SPAN
Not known

HABITAT
Bottom of tropical fresh waters, often in muddy and murky conditions

DISTRIBUTION
Basins of the Nile, Niger, Volta, Chad, Senegal and Gambia Rivers; also in Lake Rudolf, northern Kenya

STATUS
Little data available, but probably not threatened

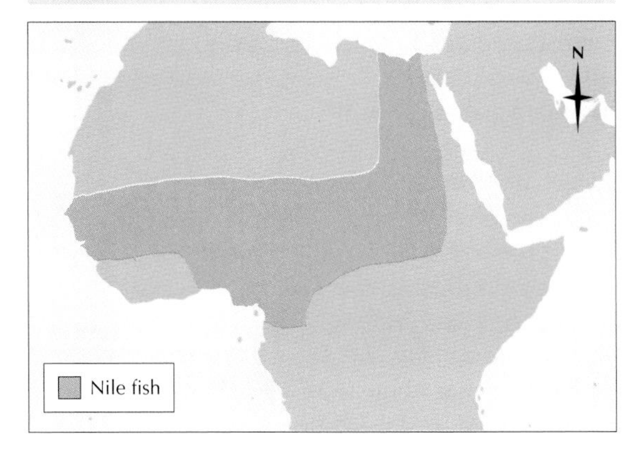

NILE PERCH

The Nile perch is the largest freshwater fish in Africa and one of the largest perch-like fish in the world. Often referred to as the king of African fish, it is valued by fishers as a game fish, and its palatable flesh makes it a popular food fish.

The Nile perch is stout-bodied, high in the back, with a relatively small head. The eyes are large and the lower jaw juts forward beyond the upper jaw. The body is covered with large scales and the gill covers bear prominent spines. The pectoral fins lie just behind the gills and immediately beneath them are the spiny pelvic fins. The dorsal fin is in two lobes, the front one having stout, spiny rays. The first three rays of the anal fin are stout and spiny and the tail fin is rounded. The adult Nile perch is dull brown, olive or gray on the back, uniformly colored or marbled, and silvery on the belly. It is believed to grow up to 6¼ feet (1.9 m) long and to have a maximum weight of about 440 pounds (200 kg).

The Nile perch ranges from Lake Albert in Uganda north to the Nile Delta, but it also lives in Lakes Chad, Rudolf and Abbaya (Democratic Republic of the Congo) and in the Senegal, Niger and Zaire Rivers. It is absent from the Zambesi and other African rivers.

A full-grown Nile perch is a very powerful fish, so much so that, when hooked, it cannot be pulled to the surface. One method used by local fishers to capture it is to tie the line to the end of the canoe and let the fish tire itself out. By day the large adults lie in water 15–20 feet (4.5–6 m) or more deep, coming into the shallows to feed at night.

Bigger and bigger meals

The Nile perch is a voracious feeder, somewhat in the manner of the pike, of the genus *Esox*. Even the young fish are cannibalistic. However, the pike and the Nile perch differ in their manner of feeding. The pike has large teeth, and opens its mouth in a "grin" when attacking prey. The Nile perch has a round mouth and large numbers of tiny, backwardly directed teeth set in broad bands. The teeth are too small to pierce the skin of prey. Instead, they form a friction pad for holding it.

Young Nile perch feed on plankton and later on freshwater prawns as well. When they are about 4 inches (10 cm) long, they start feeding on tiny fish, eating larger and larger fish as they grow bigger. Large adults seldom eat anything smaller than tiger fish, 1 foot (30 cm) in length, and medium-sized Nile perch feed also on cichlids.

In the late 1950s the Nile perch was introduced into Lake Victoria, but it had a disastrous effect on the local fish community, particularly the cichlids, and the commercial fishery there eventually collapsed. Two populations of Nile perch were introduced into reservoirs in Texas, one in 1978 and the second in 1985, intended for sport fishing. However, scientists now believe that both populations have died out completely.

Slow start in life

The adult female Nile perch is more than twice the size of the male, with an average weight of 60 pounds (27 kg), compared with an average of 25 pounds (11 kg) in the male. The eggs are small, 0.8 millimeters in diameter, with a single oil globule, enabling them to float just under the water's surface. They are only slightly heavier than water and the slightest movement in the water is enough to keep them buoyed up. Spawning takes place in relatively still waters, such as lakes, in the oxbow lakes of rivers or in flooded backwaters. The eggs hatch in less than 20 hours, so the larvae are at an early stage of

Even at 3 months, the Nile perch is a large fish. Juveniles initially feed on plankton, but in due course move on to freshwater prawns and larger fish.

The Nile perch was venerated by the ancient Egyptians. Large numbers of mummified perch have been found in the Nile Valley.

development, being hardly more than embryos and only 1.3 millimeters long. When they are ⅓ inch (7.6 mm) long, they begin to resemble normal fish larvae, with the yolk sac still attached. When ½ inch (1.3 cm) long, they begin to take on the features seen in the adults, but the body is marked with irregular dark bands.

Earliest mummified fish

The Nile perch was venerated and mummified by the ancient Egyptians. At Esneh, in Upper Egypt, it was worshipped as a divinity of first rank, so much so that the Greeks called the town Latopolis, the City of the Lates fish. There, in the valley of the Nile, at the beginning of the 20th century large numbers of the fish were dug out of the ground. They had been embalmed in brine, wrapped in cloth and tied with cords. The dry sandy soil had doubtless helped to preserve them. Although more than 2,500 years old, the flesh was in as good condition as fish cured in the sun today. Chemical analysis showed that the mummified fish contained as much animal matter as dried cod. At Gurob, about 60 miles (96 km) south of Cairo, there were other extensive burials, but these fish had been preserved in ashes or wrapped in grass and only their skeletons remained. Paintings of the Nile perch were found in Egyptian tombs. It also figured on Greek coins. Toward the end of the 19th century a bronze model of the fish, 4½ inches (11.3 cm) long, was found in Egypt. It contained the remains of a young Nile perch that had been reduced to a small pile of bones.

The extent to which ancient civilizations venerated the Nile perch is shown by the way each fish was usually buried in its own pit. If two were buried together, they were placed head to head or side by side; if more than two were placed in a grave, they were layered. The great respect with which they were held is shown by the fact that no other animal was buried in the same pit as a Nile perch.

NILE PERCH

CLASS **Osteichthyes**

ORDER **Perciformes**

FAMILY **Centropomidae**

GENUS ***Centropomus, Hypopterus, Lates* and *Psammoperca***

SPECIES **22, including *Lates niloticus* (detailed below)**

WEIGHT
Up to 440 lb. (200 kg). Average: male, 25 lb. (11 kg); female, 60 lb. (27 kg).

LENGTH
Up to 6¼ ft. (1.9 m)

DISTINCTIVE FEATURES
Stout body covered with large scales; relatively small head; prominent spines on gill cover; dorsal fin divided into 2 lobes. Adult: dull brown, olive or gray back, uniformly colored or marbled; silvery belly. Juvenile: marbled gray green.

DIET
Adult: mainly other fish; also large crustaceans and insects. Juvenile: plankton.

BREEDING
Hatching period: within 1 day

LIFE SPAN
Several years

HABITAT
Rivers, lakes and irrigation canals; adult in deep water, juvenile in shallow water

DISTRIBUTION
Major river basins of tropical Africa, including Nile, Chad, Senegal, Volta and Congo; also in Lakes Mariout, Albert, Tana, Rudolf and Abbaya

STATUS
Common

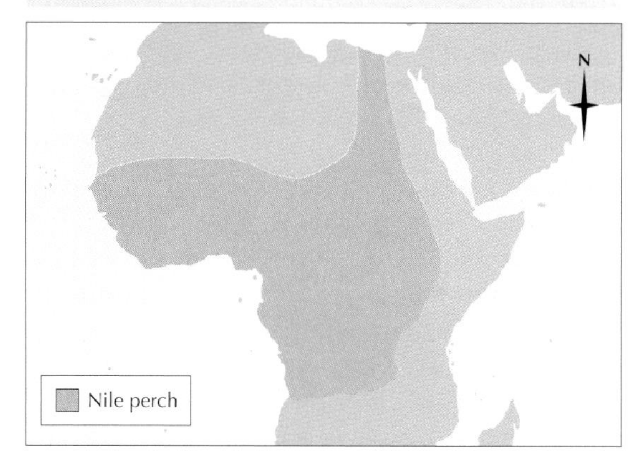

NILGAI

THE NILGAI, SOMETIMES SPELLED nylghai or nylgau, is the largest Indian "antelope," although it is not in fact a true antelope. The bull measures up to 5 feet (1.5 m) at the shoulder and may weigh 530 pounds (240 kg). The female is considerably smaller. Only the male has horns, which are very short, averaging just 8 inches (20 cm) in length. The nilgai's build is somewhat unusual, with the withers (the ridge between the shoulder bones) considerably higher than the rump. The male is steel gray in color, the female and young having a lighter, more tawny coat. An alternative name of blue buck or blue bull is derived from the blue gray coloring of the male. There is a white ring below each fetlock (a projection of the back of the leg above the hoof) and two white spots on either cheek. The nilgai has a short, stiff, black mane and the male develops a spike of dark hair on the throat.

Closely related to the nilgai is the much smaller chousingha, *Tetraceros quadricornis*, or four-horned antelope, also of India. The male is up to 2 feet (60 cm) high and 44 pounds (20 kg) in weight, with two pairs of horns. The female is similar to the male but has no horns. The coat is dull reddish brown in color and white below. There is a dark stripe down the front of each leg.

Nilgai are found in a wide range of habitats, particularly forest and grassland, but also on open plains and desert edges. They range from the base of the Himalayas in the north of India to Mysore in the south, but are not found in Bangladesh, in Assam or on the Malabar coast. They also occur in Nepal and Pakistan and have been introduced into the United States.

Creatures of habit

Nilgai are creatures of habit. Within their home ranges, the herds of four to 10 always use set places for resting, drinking and defecation. They build up large dung heaps that probably serve to mark the center of the home range and enable wandering bulls to identify if the females are in season. The herds are made up of a few cows with their young, along with one or two young bulls. The adult bulls are solitary or live in small groups except during the rut (mating season).

Nilgai feed on leaves and grass and will often drop to their knees when they are grazing, as their necks are short. Sometimes they cause considerable damage to crops. They feed mainly in the early morning and late evening, lying up in shade during the heat of the day. Although their trips to water are regular, some go for long periods without water when sources dry up.

The four-horned antelope has a distribution similar to that of the nilgai, but does not occur in Pakistan. Like the nilgai, it shelters during the heat of the day, in tall grass and open jungle, and feeds mainly in the morning and evening. Four-horned antelope live singly or in pairs, or a buck may gather four or five does around him.

Awkward but speedy

Both nilgai and four-horned antelope walk and run in a rather jerky manner. The nilgai, with its long legs, has a stilted gallop. When threatened, four-horned antelope dive into the undergrowth but nilgai herds give low alarm whistles and

Nilgai are not true antelopes, and are more closely related to cattle and buffalo.

Two female nilgai. Nilgai are generally found in small herds of up to 10 animals, often a few cows and their young. Adult bulls are mainly solitary.

gallop off slouchingly, with their heads held up. Their awkwardness is, however, deceptive. They run as fast as a horse and can go over much rougher ground. Nevertheless, they are preyed on by leopards, tigers, wolves and dholes but are rarely molested by humans because they are closely related to cows, which are held sacred. Both species are easily tamed when young.

Males wrestle for females

During the rut the males compete to join the females' herds. Two bulls will drop to their knees, pressing their foreheads together, and wrestle, or push with their necks. Although breeding takes place throughout the year, the usual mating time is in the spring. The one or two young are born 243–247 days later, in November and December, and are sexually mature at 3 years in the case of females and 5 years in males.

Of ancient stock

Both the nilgai and four-horned antelope belong to the "bovine" group of antelopes, closely related to cattle and buffalo. Their nearest relatives among antelopes are the African bushbuck, kudu and eland. They are thought to be very primitive, having hardly changed since Miocene times, 15 million years ago.

Unlike the bushbuck and its relatives, the nilgai and four-horned antelope share some important skull characteristics with cattle and buffalo. They have a long palate, their molar teeth have extra pillars on them and high crowns, and they have symmetrically shovel-shaped incisors. Also, they lack the elaborately twisted horns of bushbuck and eland antelopes.

NILGAI

CLASS **Mammalia**

ORDER **Artiodactyla**

FAMILY **Bovidae**

GENUS AND SPECIES ***Boselaphus tragocamelus***

ALTERNATIVE NAMES
Blue bull; blue buck

WEIGHT
Male: up to 530 lb. (240 kg); female: up to 375 lb. (170 kg)

LENGTH
Head and body: 6–7 ft. (1.8–2.1 m); shoulder height: 4–5 ft. (1.2–1.5 m); tail: 1½–1¾ ft. (45–54 cm)

DISTINCTIVE FEATURES
Short body hair; stiff, dark mane; shoulders higher than rump. Male: steel or blue gray; paler underside; spike of dark hair on throat; very short horns. Female: tawny upperparts; lacks horns.

DIET
Mainly leaves and grasses; also crops

BREEDING
Age at first breeding: 3 years (female), 5 years (male); breeding season: all year, with peak in spring; number of young: 1 or 2; gestation period: 243–247 days; breeding interval: 1 year

LIFE SPAN
Up to 15 years

HABITAT
Mainly forest and grassland

DISTRIBUTION
India, Nepal and Pakistan; introduced to parts of U.S.

STATUS
Relatively common, but dependent on conservation programs

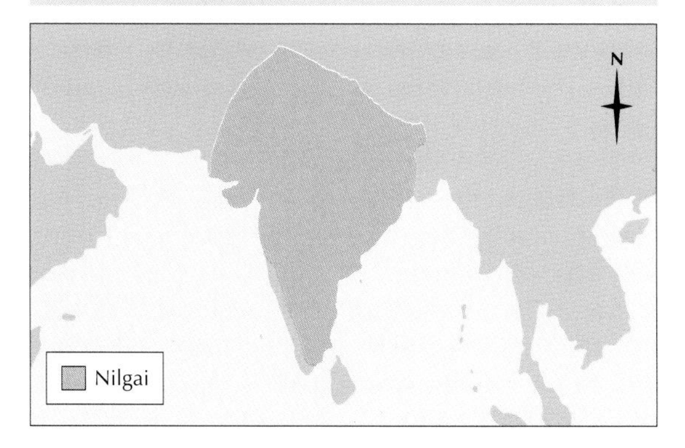

NODDY

In common with other terns, noddies are delicately built birds 12–16 inches (30–40.5 cm) long with slender wings, a pointed bill and short legs. The name noddy comes from the birds' habit of nodding and bowing to each other during courtship.

The brown noddy (above) breeds on tropical and subtropical islands, and feeds over the open ocean.

There are five species of noddies. The black noddy, *Anous minutus*, is 14–15 inches (35–39 cm) long with dark brown, often nearly black plumage and a pale, almost white cap running from the base of the bill over the top of the eyes. It has a distinct white ring under the eyes. The brown noddy, *A. stolidus*, is the largest of the species, growing up to 16 inches (40.5 cm). It breeds on tropical and subtropical islands in the Atlantic, Pacific and Indian Oceans and has similar plumage to the black noddy. The other noddies have lighter plumage. The gray noddy, *A. albivittus*, is almost entirely gray except for black rings around the eyes. It is confined to the Pacific Ocean from Hawaii to Easter Island, while the other noddies are found in most tropical and subtropical seas. The blue noddy, *Procelsterna cerulea*, has a pale gray-blue plumage, and the lesser noddy, *A. tenuirostris*, is similar in appearance to the brown noddy, but is much smaller, with a slimmer and relatively longer bill. It is confined to the tropical Indian Ocean and the west coast of Australia.

Noddies spend much of their time at sea and usually nest on cliffs on fairly remote islands. Between 1957 and 1959 the black and brown noddies were studied by members of a British Ornithologists' Union Expedition to Ascension Island in the South Atlantic. The ornithologists could reach only a few noddy nests to ring the occupants, but they were able to observe events at the nests. These two noddy species are of particular interest because of their cliff-nesting habits. By contrast, other terns mostly nest on flat ground. Elsewhere, noddies nest in trees and bushes, while in Belize, black noddies nest in mangroves and on the ground. On Ascension Island the noddies occupy nesting ledges year-round, but egg clutches are laid early in the year.

Swooping for fish

Unlike their relatives, noddies do not plunge-dive, but feed on small, surface-living fish. They swoop from about 20 feet (6 m) and catch the fish or squid when they are just under the surface. Anchovies, flying fish and squid form the bulk of the noddies' diet. Quite often noddies feed with seabirds that dive, catching the fish as they come to the surface or even as they leap into the air to try to escape from the other birds. They have also been seen congregating over shoals of tuna as these predatory fish force smaller fish to take desperate measures to avoid being eaten. Noddies generally catch fish that are about 3 inches (7.5 cm) long but they sometimes bring prey as large as 7 inches (17.5 cm) long back to their waiting hungry chicks.

The brown noddy feeds up to about 30 miles (50 km) from its breeding colony. It often drinks or bathes in flight, both during the day and by moonlight. Noddies regularly perch on flotsam, buoys and ships' rigging and sometimes on the backs of turtles, or even swimming pelicans.

Competition for space

When noddies nest on cliffs there is often competition for ledges on which the eggs and chicks can be reared, even when there seem to be many suitable ledges on the cliffs. This is because noddies prefer to nest close together. As a result in each colony some pairs have to make do with poor sites and risk losing their eggs or chicks. The competition for nesting space is somewhat eased by the gradual accumulation of guano on the ledges and by the efforts of the noddies

An adult lesser noddy on its nest, Seychelles, Indian Ocean.

themselves in building nests of guano bound by feathers, seaweed or leaves, although such artificially widened ledges are not completely safe.

Noddies lay a single egg, whereas most other terns lay two or three. Incubation takes about 5 weeks and is shared by both parents, which take turns incubating the eggs for 1–2 days each. The chicks hatch with a covering of down. Those of black noddies have the same pattern of dark brown and white as their parents, but those of brown noddies may be one of a variety of colors ranging from dirty white to brownish black. The adults feed the chicks on fish and squid, which they pass directly to them rather than dropping the food onto the ledge. The chicks fly when they are 7–9 weeks old. Those that grow up on ledges have to fly well from the start, but if the chicks are raised on bushes or fairly flat ground they are able to make practice flights.

The chicks belonging to early broods on Ascension Island stand a better chance of survival than those hatched later because there is a shortage of food there in the fall. Post-fledging feeding by the parents lasts for several weeks. The chicks are fed regurgitated fish or whole fish as they grow.

Safety on ledges

When they nest in bushes or on the ground, noddies sometimes fall prey to cats or rats that have been introduced into their island homes. However, they are safe on the cliff ledges, where even predatory frigate birds are unable to reach them because of the difficulty of landing on a narrow ledge. Noddies swoop at intruders, stabbing at them with their bills, to deter attack.

BLACK NODDY

CLASS **Aves**

ORDER **Charadriiformes**

FAMILY **Sternidae**

GENUS AND SPECIES ***Anous minutus***

WEIGHT
3⅓–3⅘ oz. (95–110 g)

LENGTH
Head to tail: 14–15 in. (35–39 cm); wingspan: 26–28 in. (66–72 cm)

DISTINCTIVE FEATURES
Delicate body; long, slender wings; long, thin, pointed bill; short black legs and feet; grayish white forehead, nape and crown; white ring under eye; long grayish tail with forked notch at tip; rest of plumage brownish black

DIET
Mainly small fish; prey changes according to season

BREEDING
Age at first breeding: probably 1 or 2 years; breeding season: year-round; number of eggs: 1; incubation period: about 36 days; fledging period: 48–60 days; breeding interval: 1 year

LIFE SPAN
Up to 18 years

HABITAT
Open seas and oceans; nests on cliffs, trees, bushes or the ground, usually on small, remote islands

DISTRIBUTION
Tropical Pacific, Atlantic and Indian Oceans; also in Caribbean

STATUS
Locally common

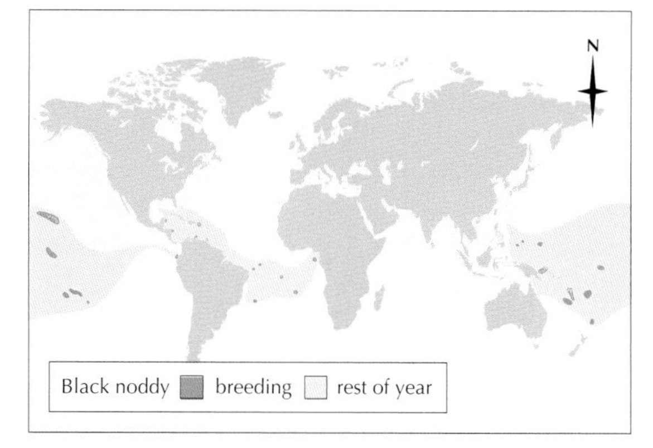

NUTCRACKER

Clark's nutcracker (above) has similar habits to the European nutcracker, but its plumage is very different to that of its spotted relative.

THE TWO SPECIES OF nutcrackers are members of the crow family. The European nutcracker, *Nucifraga caryocatactes*, grows to about 12 inches (30 cm) long, and is dark brown with white flecks. It has broad wings, a short tail and a long, pointed bill. The European nutcracker ranges across central and northern Europe, but in European Russia and northern and Central Asia there is a subspecies, the slender-billed nutcracker, *N. c. caryocatactes*. The thick-billed and slender-billed subspecies are difficult to tell apart in the field. The latter sometimes ranges into Europe, coming as far west as Britain. Clark's nutcracker, *N. columbiana*, lives in western North America and has similar habits to its European relative. Its tail and wings are black and white, and the rest of its plumage is gray.

Jaylike in habits

Nutcrackers are at home in pine forests or mixed woodlands in which conifers are plentiful. Outside the breeding season they disperse more frequently into deciduous woods, especially where there is hazel. Then they move about in loose parties, spending much time on the ground, hopping heavily as jays do. They also fly over the tops of tall trees or perch on the highest twigs. Their calls are harsh but less strident than those of jays and have a greater carrying power. In spring their call is an almost musical babbling but they are always silent in the nesting season.

Basic diet of pine seeds

Inside the nutcracker's bill is a projection in the lower part that fits into a cavity in the upper part. Together these adaptations form a highly efficient "nutcracker." The bird's food is largely the seeds from pine, spruce, cedar and larch cones. The European nutcracker is particularly attracted to the large wingless seeds of the Cembran or Arolla pine, also called the Swiss stone pine, *Pinus cembra*. The seeds are usually picked out of the cones while they are hanging on the tree, but the nutcracker may eat by holding the cones with its feet while sitting on a branch or on the ground. This is particularly true of the Arolla pine. The nutcracker also eats acorns, beechmast, hazelnuts and walnuts as well as juniper berries. Insects and earthworms are taken, as are the eggs and young of small birds. Conifers supply the basic foods for the nutcracker's diet, however, and this determines

A European nutcracker with a hazelnut. The nutcracker is named for its specialized bill, which enables the bird to pierce nutshells with ease.

the bird's distribution and breeding range. Nutcracker movements are influenced by the crop of conifer seeds, which varies from year to year.

The link between the Arolla pine and European nutcrackers is so strong that some scientists believe the relationship is symbiotic (of mutual benefit). The nutcrackers hoard the seeds by pushing them into the ground or into rocky crevices, where they lodge and germinate far more successfully than seeds that fall naturally from the cones and lie on top of the ground. In this way the nutcrackers' actions provide the heavy and unwinged Arolla seeds with a more effective means of dispersal than they would otherwise achieve.

Young fed on hazelnuts

European nutcrackers always nest in a conifer, 15–30 feet (4.6–9.1 m) from the ground, usually near the trunk. The nest itself is made of twigs, moss and lichens, reinforced with earth and lined with grass and the hairy lichen, *Usnea barbata*. Usually a clutch consists of three or four eggs. These are bluish green with olive brown or gray markings, and are laid during March and April. They are incubated for 18 days by the female, which is fed by her mate. The young remain in the nest for 3–4 weeks. Both parents feed them with hazelnuts buried the previous fall, which they carry to the nest in their throat pouches.

EUROPEAN NUTCRACKER

CLASS **Aves**

ORDER **Passeriformes**

FAMILY **Corvidae**

GENUS AND SPECIES ***Nucifraga caryocatactes***

WEIGHT
5–6¾ oz. (140–190 g)

LENGTH
Head to tail: 12½–13 in. (32–33 cm); wingspan: 20½–23 in. (52–58 cm)

DISTINCTIVE FEATURES
Long, powerful bill with sharp point; broad wings; rather short tail; chocolate-brown plumage with heavy white spotting almost all over; pure white undertail area and corners to tail

DIET
Mainly seeds and nuts, particularly seeds of conifer trees; also invertebrates and occasionally bird eggs and nestlings

BREEDING
Age at first breeding: 1 year; breeding season: eggs laid March–April; number of eggs: usually 3 or 4; incubation period: about 18 days; fledging period: 24–25 days; breeding interval: 1 year

LIFE SPAN
Up to 8 years

HABITAT
Forests, especially with Arolla pine, Siberian stone pine and Norway spruce; also hazel woodland

DISTRIBUTION
Parts of central and northern Europe, east to Central Asia, Siberia, China, Korea and Japan

STATUS
Common

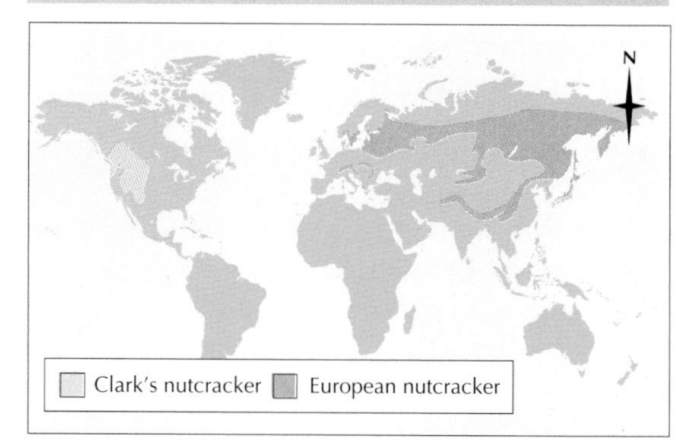

Nutcrackers always cover their caches (food stores) after placement, usually with a leaf, stick or stone. In the depths of winter, when food is in short supply, nutcrackers unerringly dig through the snow to find the caches. Recent research suggests that some birds that store food in the fall experience an expansion of part of the brain to help them remember where food is buried. Scientists now believe that a seasonal increase in an area of the brain called the hippocampus allows birds to find their caches. It could be that this applies to nutcrackers also.

The success or otherwise of the birds' breeding is directly linked with the hazelnut crop. When this is good, as many as four or five eggs will be laid and there will be more young birds surviving the winter to breed the following year.

Migrations and irruptions

The areas in which thick-billed nutcrackers breed in central and northern Europe are limited largely to the mountains of Switzerland, the Carpathians and Balkans, and to southern Scandinavia. The slender-billed nutcrackers, less tied to mountain forests than their thick-billed relatives, extend from the forests of Baltic Russia across Siberia to Korea and Japan. There is, however, an isolated breeding area in Taiwan and a more extensive zone through the foothills of the Himalayas and into southern China.

The slender-billed race regularly experiences irruptions (sudden upsurges in the population numbers of a particular area, often brought about by disturbances in natural ecological balances). These irruptions are associated with the failure of the pine seed crop, especially that of the Siberian stone pine. Little or no return migration takes place the following spring, with most birds failing to survive. In the last 40 years irruptions from Siberia have brought birds to western Europe in 1968, 1971, 1977, 1985 and 1995. Sometimes new breeding colonies are established in areas invaded in these movements.

Clark's nutcracker has a similar diet to that of the European nutcracker, though the species of pine it feeds on are different. A Clark's nutcracker may store up to 30,000 pinyon pine seeds in a single season, placing them in caches of four to five seeds each. Like its European relative, Clark's nutcracker has to adapt to fluctuations in the annual seed crop, and in poor years it moves down to lower, warmer altitudes in the fall and winter, such as coastal and desert regions. This movement occurs every 10 or 15 years in the western United States. Since 1898 there have been several spectacular irruptions, including the years 1898–1899, 1919–1920, 1935–1936, 1950–1951 and 1955–1956. These figures reveal that, as in the European nutcracker, there are no regular irruption cycles for Clark's nutcracker.

Clark's nutcrackers inhabit high coniferous forest in the western United States. About every 10 or 15 years they irrupt into desert and lowland areas in the west.

NUTHATCH

The Eurasian nuthatch feeds on hazelnuts in the fall, which it opens by jamming them into a crack in tree bark and hammering at them with its bill.

NUTHATCHES ARE SMALL birds 4–7½ inches (10–19 cm) long, with compact bodies, short tails, long, strong toes and claws and long, tapering bills. Generally they are bluish gray above and white, gray or chestnut on the underparts, often with a black stripe across the eye. There are 24 species of true nuthatches, four of which are found in North America.

The most common and widely distributed species is the Eurasian nuthatch, *Sitta europaea*, which ranges across much of Europe and Asia except for the extreme north. It is 5½ inches (14 cm) long, slate gray on the upperparts, and buff underneath with chestnut-red flanks. There is a conspicuous black stripe across the eye, and the cheeks and throat are white. The tail is stumpy, black at the tip with white markings on the outer feathers.

The Eurasian nuthatch lives in woods, parkland and large gardens with old broad-leaved trees. In Russia and eastern Asia it may live in dry pine forests high up the mountains although in much of the rest of Europe it is only occasionally found in pinewoods. The Eurasian nuthatch is one of the noisiest woodland birds, and it advertises its presence by a loud, frequently repeated, metallic call: *chwit-chwit*.

A nuthatch leaps in short jerks over the trunks and branches of trees, upward, downward or sideways with equal ease. It does not use its tail as a support when it climbs, as woodpeckers and tree creepers do, and holds on with only its clawed toes. Two toes on each foot face forward and two backward, enabling the bird to grip strongly onto the bark of trees. When a nuthatch rests, it clings with one foot and props itself by placing the other one lower down on the tree. When it goes to roost, in a depression or a crevice in the bark, the nuthatch settles itself head downward. The ability to move easily both upward and downward, headfirst, on a tree trunk makes the nuthatch unique among birds.

Hammer blows

The Eurasian nuthatch was originally called the nuthack. This name derives from the fact that in the fall its main food is hazelnuts. To consume them, it first places the nuts in crevices in the

EURASIAN NUTHATCH

CLASS **Aves**

ORDER **Passeriformes**

FAMILY **Sittidae**

GENUS AND SPECIES ***Sitta europaea***

WEIGHT
¾–1 oz. (21–26 g)

LENGTH
Head to tail: about 5½ in. (14 cm); wingspan: 9–10¾ in. (22.5–27 cm)

DISTINCTIVE FEATURES
Compact body; stout, pointed bill; short, yellowish brown legs with strong claws; rather short tail; blue-gray upperparts; white throat merging into orange-buff underparts; black stripe across eye from bill to sides of neck

DIET
Invertebrates and seeds; also hazelnuts when in season

BREEDING
Age at first breeding: 1 year; breeding season: eggs laid April–May; number of eggs: 6 to 11; incubation period: 13–18 days; fledging period: 23–24 days; breeding interval: 1 year

LIFE SPAN
Up to 9 years

HABITAT
Deciduous, mixed and coniferous woods; also parkland and large gardens with mature trees

DISTRIBUTION
Much of Europe, south to northwest Africa, east through Middle East and Central Asia to Japan, India and Southeast Asia

STATUS
Common

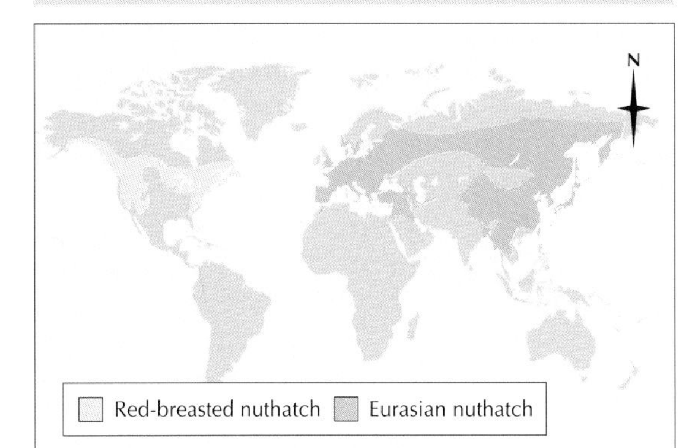

The red-breasted nuthatch favors coniferous forests. Its numbers and winter range vary according to the crop of conifer seeds each year.

bark low down on the trunks of mature trees such as oaks. Then it hammers the nut with its bill, swinging its body with each blow. The half shells usually remain fixed in the bark or fall to the base of the tree. In common with other nuthatch species native to northern regions or mountainous areas, the Eurasian nuthatch also feeds on beechmast, acorns, the seeds of yew and similar seeds, especially in winter and the fall. Nuthatches search the crevices of bark at other times during the year for insects, including beetles, earwigs, flies and bugs, and they open galls (swellings in plants) to extract grubs. They also feed on spiders, small snails and other small invertebrate animals.

With bird feeders becoming more and more popular in many gardens, nuthatches have tended increasingly to come near houses to take bread and fat from the feeders, often wedging the food in crevices in trees. There is at least one record of a nuthatch actually burying a nut,

Nuthatches are the only birds that can climb both up and down trees headfirst. Pictured is a white-breasted nuthatch, **Sitta carolinensis.**

pushing it deep down between the grass and the base of a tree with hard thrusts of the bill, and then covering the hiding place with grass.

Plastering with mud

Eurasian nuthatches breed in April or May, choosing a cavity in the trunk of a tree or a stout branch, or sometimes a hole in a wall or in a nesting box. The entrance is made smaller by the nuthatches plastering mud around the opening, which becomes very hard when it is set. The cavity chosen may be an old woodpecker nest or a natural rot cavity and sometimes a bank swallow's hole is used. The nest inside is made of dead leaves, bits of bark or grass, and both partners share the work of constructing it. The eggs, white and usually spotted with brown and streaked with violet, may number six to eleven, with occasionally only four or five. Incubation, which begins after the last egg is laid, is by the female only and lasts 13–18 days. The young are fed by both parents, initially on small insects and later on caterpillars and spiders.

A high mortality rate?

There is little information about the predators of nuthatches, but because the number of eggs in a clutch is high ornithologists believe that there must be a high death rate among the young. These birds are no more vulnerable to birds of prey or ground predators than other small birds, and their nests probably provide greater protection than most. There are indications that the small opening to the nest may constitute a danger to the young birds when making their first exit, as it may cause them to inadvertently damage their legs or wings. If this is true, it could indicate that there is an unusually high death rate among these young birds.

Canada–Corsica axis

The red-breasted nuthatch, *S. canadensis*, is native to North America. It has a white stripe above a black eye stripe, a black cap to the head and rust-colored underparts. This species extends across southern Canada and southward into the western United States and into the New England states in the east. It is one of the few nuthatch species known to be migratory and in the winter withdraws from higher altitudes and mountainous areas. The Corsican nuthatch, *S. whiteheadi*, is virtually indistinguishable from the red-breasted nuthatch. It lives on the Mediterranean island of Corsica and nowhere else in Europe. The Corsican nuthatch gathers seeds from the fall onward, storing them under moss or in crevices in bark, to be eaten later.

Very similar-looking species of nuthatch live in eastern China and the Middle East. Considering how far apart these populations are located, the similarity in appearance is surprising, yet the species must be closely related. After making a close study of nuthatches from these areas, leading ornithologists concluded that the birds belong either to one species or to four virtually indistinguishable species.

NUTSHELL

ALTHOUGH INSIGNIFICANT compared with some of the larger, more colorful marine shells, nutshells are interesting because of their unusual ways of feeding and moving about. They are also very numerous.

The nutshell is a primitive, bivalve mollusk shaped rather like a hazelnut, rounded but nearly triangular in outline. The many species, which all look much alike, are found in all temperate and tropical seas, at depths of between 33 and 660 feet (10–200 m). Nutshells are mainly small, usually about ½ inch (1.3 cm) long, but may be up to ¾ inch (2 cm) and occasionally larger. The shell is white or gray in color with patches of olive brown and slightly coarser concentric rings. There are often reddish brown or purple-gray rays or lines radiating from the hinge region. The color is mainly in the outer protein layer, but this may wear off empty shells. At the apex of the shell are the hinge and the dark brown elastic ligament that pushes the shell open. On either side of this, each valve has a row of teeth that interlock and help to keep the valves aligned. The edge of each valve is notched except in the species known as the thin nutshell, *Nuculus tenuis*..

Nutshells are one of the most primitive of bivalves. Their immediate ancestors lived nearly 500 million years ago, while nutshells not unlike those living today were on Earth nearly 400 million years ago. The thin nutshell lives off all the coasts of the Northern Hemisphere.

The muddy life

Nutshells can be so common in some areas that they are the most abundant of the invertebrates, apart from those of near-microscopic size. They live on the seabed, wherever they can bury themselves, each species thriving best on deposits of a particular degree of coarseness, from mud to sandy gravel. They burrow using a muscular foot. This is unlike that of most other bivalves in that it ends in a sole that is folded down the middle, with the two halves facing outward. The foot is thrust out from between the valves and the two halves are opened at the tip to grip the seabed. The rest of the foot then contracts and draws the body along.

There really is little need for nutshells to move, however, because they feed on deposits, their food being made up of small organic

Nutshells are primitive bivalve mollusks. They are unusual in that they actively probe for food particles using a muscular proboscis.

particles in the sand or mud with which the animal is normally barely covered, hinge uppermost. A current of water for respiration, containing minute food particles of mainly dead animal and plant matter, is drawn in at the front of the shell by beating cilia (hairlike, vibrating organs) and then driven out at the rear. Sometimes, when too much debris accumulates, the valves will suddenly close. This drives a strong current across the gills to clean them. In addition, siphons extend to the surface of the mud or sand where they ingest particles of food.

Probing for food

Most bivalves feed by filtering small particles from the water using their gills. The particles are passed for sorting to smaller, ciliated flaps, called palps, that direct food particles to the mouth. In the nutshell the gills are small and simple, while the palps are large and complex and have something not found in other bivalves: a pair of long palp-proboscides. At the front end of each palp is a proboscis long enough to be pushed some way out of the shell. It is very muscular, contractile and active. Along one side is a groove lined with cilia, and as the proboscis feels around in the sand or mud, small organic particles are picked up and carried along the groove to the main part of the palp via a flaplike structure called the palp pouch. As well as the particles brought in by the palp-proboscides, there are the others carried in with the water current. Some of these are caught on the outer surfaces of the palps and more are passed from the gills to the palps in the same way as in other bivalves.

Hermaphroditic breeding

Nutshells are hermaphrodites; they have both male and female reproductive organs. The breeding season varies from species to species. Eggs, about 0.1 millimeters across, and sperm are shed directly into the sea through the kidneys. After fertilization, the egg develops into a larva that swims by the beating of three belts of cilia that surround its barrel-shaped body. The ciliated larvae remain in the plankton for a few weeks. It is these planktonic larvae that distribute the otherwise sedentary nutshells. Eventually, the outer layer of cells of each larva is cast off to reveal a young bivalve complete with a tiny shell. At this point the nutshells will move down to settle into the mud or sand. It is not known exactly how long nutshells live, but they are certainly capable of living for 3 or 4 years.

This is the typical sequence of development for a nutshell species, but there are some interesting departures from it. For example, in an American species, *Nucula delphinodonta,* the larvae are kept for 3 or 4 weeks in a brood sac of mucuslike material mixed with foreign bodies. This is attached to the hind end of the shell. With this nourishment and protection of the young, each larva has a much increased chance of survival. As a result, only 20 to 70 eggs are laid, these being larger than usual.

NUTSHELLS

PHYLUM **Mollusca**

CLASS **Bivalvia**

ORDER **Protobranchia**

FAMILY **Nuculidae**

GENUS AND SPECIES **Brown nutshell, *Nuculus sulcata*; thin nutshell, *N. tenuis*; many others**

LENGTH
Up to ¾ in. (2 cm), sometimes larger

DISTINCTIVE FEATURES
Rounded bivalve mollusk. White or gray shell with patches of olive brown; thin, radiating lines on shell with slightly coarser concentric rings.

DIET
Organic particles of food from surface of mud or sand

BREEDING
Hermaphroditic: each nutshell releases eggs and sperm directly into sea. Breeding season and number of eggs vary with species.

LIFE SPAN
At least 3–4 years

HABITAT
In mud and sand at depths of 33–660 ft. (10–200 m)

DISTRIBUTION
Coasts of eastern Atlantic; Mediterranean Sea

STATUS
Very common

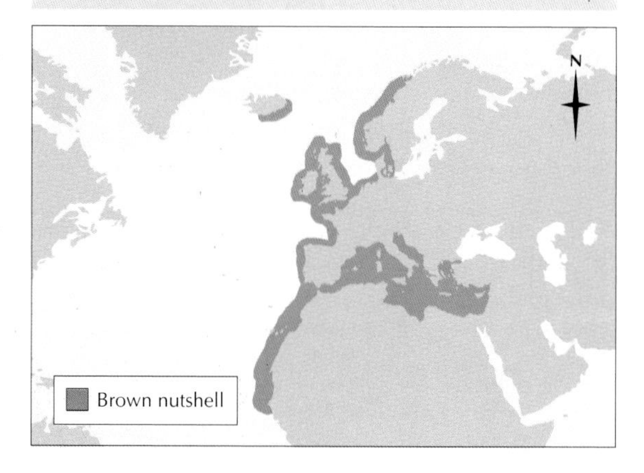

NYALA

THE NYALA AND ITS RELATIVES form a group of oxlike antelopes belonging to the genus *Tragelaphus*, including the bushbuck (*T. scriptus*), bongo (*T. eurycerus*), sitatunga, (*T. spekeii*), greater kudu (*T. strepsiceros*) and lesser kudu (*T. imberbis*). These species are discussed in separate articles. All have white spots and stripes on the body and throat fringes, and the males have twisted horns, some with yellow tips. Nyala occur in northeastern Natal, eastern Transvaal, Zimbabwe and southern Malawi.

The true nyala or inyala, *T. angasi*, which stands 32–48 inches (0.8–1.2 m) high and weighs 155–285 pounds (70–130 kg), is a very striking antelope. The male, quite different in appearance from the female, is slate gray with a red forehead and a long fringe of hair on the throat and underparts. The upper lip, chin, ear bases and two cheek spots are white, and there is a white stripe across the nose, a white chest band and about 14 pale white body stripes. A few haunch spots, the dorsal crest and the underside of the tail, are also white. The legs, from just above the knees and hocks (the area above the foot on the hind limb) to the hooves, are bright tan with a white patch on the inner sides of the knees and hocks. The female is bright chestnut, with a white face band, about 11 prominent body stripes and fewer white haunch spots. Her dorsal crest is shorter than the male's and is only white where the stripes cross it. She has no throat fringe.

The mountain nyala, *T. buxtoni*, is 36–54 inches (0.9–1.35 m) high and weighs 330–660 pounds (150–300 kg). It has a longer, coarser coat than the nyala, and a shorter throat fringe. It is brownish gray, with a short dark mane and no body stripes, but it has a characteristic line of nine spots along the haunches. As well as face marks and a chest patch, it has a crescent on the throat. The inner sides of the forelegs are white.

A female nyala with her young. Nyala browse and shelter in thickets during the day, moving out into more open spaces at night.

Keeping well hidden

Nyala keep to the thickets, scrub and woodland of southeastern Africa and they are shy, secretive animals. However, they have now grown accustomed to tourists in the National Parks and tend to come out more frequently into the open.

Nyala usually live in lush, green, river country, where they feed on grasses, leaves and fallen fruits. The males live singly or in small groups. The females and young form separate herds of 8 to 16, which commonly associate with impala and eland herds where they are often unnoticed because, except for the stripes, they are much the same color. They constantly wag their tails from side to side as they move about. Like kudu, nyala have hairy glands on their hind feet that leave their scent where they walk. Mountain nyala and sitatunga do not have these. When a male nyala displays, for example when entering the females' herds during the rut (mating season), he raises his white dorsal crest, lowers his horns and moves stiffly.

Male nyala are normally peaceful but become very fierce when they are attacked. Predators include leopards and sometimes lions and Cape hunting dogs or African wild dogs.

Gestation in the nyala is thought to take 220–270 days, and a single offspring is born, usually in September or October. The young lie hidden in grass until they are old enough to

Male nyala are sexually mature after 18 months but rarely mate until about 3 years of age, by which time they have long horns, a dark coat, a crest and hair fringes.

move around with the rest of the herd. The gestation of the sitatunga is nearer 8 months, and that of the bongo about 9 months.

Mountain refuge

The mountain nyala of Ethiopia is rarer than its southern relative, the nyala. The mountain nyala's horns form a fairly open spiral featuring 1¼ turns. The mountain nyala lives high up in the mountains of the Ethiopian plateau, in the high moorlands among the giant heaths and St. John's wort east of the Rift Valley. It lives at altitudes of 9,500–12,500 feet (2,895–3,810 m) or even higher, from Mount Gugu southwest to the Chilalo and Kaka mountain ranges and the Bale Mountains. Mating in this species occurs in December and the young are born 8–9 months later toward the end of the wet season.

In 1963 the mountain nyala's numbers were low, perhaps only about 2,000 in all, although at the time it was not thought to be in immediate danger. Four years later the population was estimated again and this time calculated to be 4,000 to 5,000. Today this figure is nearer 3,000 and the species is regarded as endangered.

NYALA

CLASS **Mammalia**

ORDER **Artiodactyla**

FAMILY **Bovidae**

GENUS AND SPECIES **Nyala, *Tragelaphus angasi*; mountain nyala, *T. buxtoni***

ALTERNATIVE NAME
***T. angasi*: inyala**

WEIGHT
***T. angasi*: 155–285 lb. (70–130 kg)**

LENGTH
***T. angasi*. Head and body: 54–78 in. (1.35–1.95 m); shoulder height: 32–48 in. (0.8–1.2 m); tail: 14½–22 in. (36–55 cm).**

DISTINCTIVE FEATURES
***T. angasi*. Male: shaggy brown, chocolate or gray hair; crest of erect white or gray hair along spine; paler underparts and inner thighs; spiral horns; pale white body stripes. Female: bright chestnut; prominent white body stripes; no horns or shaggy hair.**

DIET
Mainly grasses, leaves and fallen fruits

BREEDING
***T. angasi*. Age at first breeding: 1–2 years; breeding season: all year; number of young: 1; gestation period: 220–270 days; breeding interval: 1 year.**

LIFE SPAN
Up to 19 years

HABITAT
Savanna thickets and woodland

DISTRIBUTION
***T. angasi*: southeastern Africa.**
***T. buxtoni*: Ethiopian highlands.**

STATUS
***T. angasi*: dependent on conservation programs. *T. buxtoni*: endangered.**

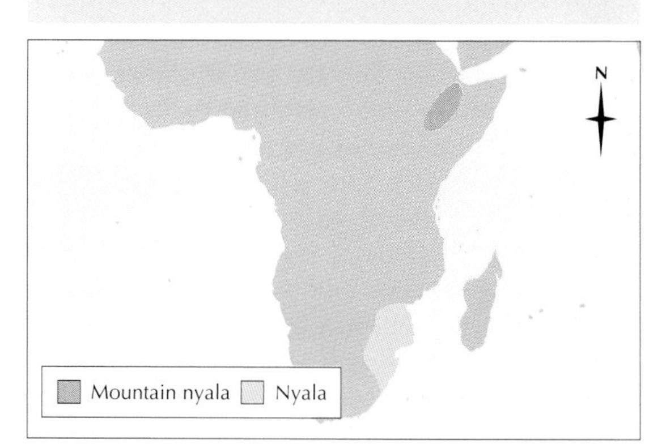

OARFISH

The oarfish has been described as one of the most mysterious of the larger sea fish and has also often been linked with legends and tales of sea serpents. Its real interest, however, lies in its unusual shape, its large size and how little we know of its way of life. There are three species of oarfish, belonging to three different genera. The best known of these is the king of the herrings, *Regalecus glesne,* so named because it was once believed that it swam in front of the herring shoals, as if leading them. Although mainly restricted to the ocean depths, oarfish large and small have been found floating on the surface or washed ashore in warm and temperate seas worldwide. The other two species of oarfish are *Agrostichthys parkeri* and *Gymnetrus russelii.*

***Elongated rays at the front of the dorsal fin make a striking mane or crest over the head. Pictured is a preserved specimen of the king of the herrings,* R. glesne.**

The longest fish

Oarfish are reputedly the longest of all fish and are listed in many authoritative publications as the longest bony fish. They have flattened, ribbonlike bodies up to 1 foot (30 cm) deep and only 2 inches (5 cm) across. The king of the herrings has been reliably documented to grow up to 36 feet (11 m) in total length. However, much larger specimens, of up to 56 feet (17 m), have also been reported on several occasions.

Hard knobs stud the skin of the oarfish, which is silvery with a bluish tinge on the head and is marked with dark, wavy streaks and sometimes with dark spots or blotches. The oarfish has a concave head profile, large eyes and a highly protrusible mouth, a characteristic of the order Lampridiformes. Its fins are pink or coral red. The dorsal fin starts on top of the head between the eyes and runs all the way to the hind end of the body. Because the front rays of the dorsal fin are long, they make a striking mane or crest over the head. The tail fin is very small or missing altogether. There are tiny spines projecting laterally off each tail and pelvic fin ray. The pectoral fins are small, but the pelvic fins, lying just under them, are long and slender, broadening at their tips, like oars. It is from the elongated pelvic fins that the oarfish gets its name.

Live at great depths

In the past varying opinions put the oarfish's habitat anywhere between the ocean's surface to depths of 3,000 feet (915 m) or more. However, these fish are now known to be pelagic (occurring in the open sea) and able to live to depths of 3,280 feet (1,000 m). The king of the herrings is widely distributed in the Atlantic, including the Mediterranean, in the Indian Ocean and in the eastern Pacific from Topanga Beach in southern California south to Chile.

It used to be the case that oarfish were rarely reported because they are so seldom seen at the water's surface. It is only quite recently that adult oarfish have been caught in nets, possibly because of the speed at which they can slip through the water with wavelike movements of their long, thin bodies. However, they are now caught with encircling nets and are marketed fresh. Rarely seen in the past, oarfish are now thought to be fairly common.

Large fish, small prey

Oarfish have very small mouths and no teeth. The king of herrings species has a large number of long, spiny gill rakers, the number varying from 42 to 58. These are used to strain very small crustaceans, especially those known as euphausids, from the water passing over the gills. Oarfish also feed on small fish and squid.

Elongated streamers

Breeding habits are little understood in the oarfish. Spawning takes place between July and December and the larvae, which hatch from small, floating eggs, may be seen near the water's surface at this time. The larvae have remarkably long streamers, ornamented with small tags of skin. These streamers are made up of the much elongated rays of the front part of the dorsal fin and the similarly elongated pelvic fins. What purpose they serve is not known.

Survives tail loss

There is no direct evidence about predators on the oarfish. Nonetheless, a high percentage of captured oarfish have either lost a part of the tail or have scars from old wounds somewhere on the rear half of their bodies. An oarfish's internal organs are all packed into the front quarter of its body. However, there is a large bag connected with the stomach, an accessory digestive organ, which extends back among the muscles of the tail to about the center of the body. Therefore, or so it seems from studying the captured specimens, an oarfish can survive an attack by a predator such as a shark, provided that only its tail or the rear half of its body is bitten. It is thought that the oarfish might still be able to survive even if its rear half is bitten off completely in an attack.

Regal coincidence

Related to the oarfish are the dealfish, family Trachipteridae. These fish have a body shape that is similar to that of the oarfish, albeit shorter and more stout. They also lack the "mane" and have only small pelvic fins. Dealfish have a small, fan-shaped tail fin, which points obliquely upward. They probably grow to 8 feet (2.4 m) long and there are eight species belonging to three different genera. The dealfish, *Trachipterus arcticus,* for example, lives in the North Atlantic while a second species, the ribbonfish, *T. trachypterus,* is found in the Mediterranean.

Another species, the whiptail ribbonfish, *Desmodema lorum,* is sometimes seen off the Pacific coast of North America, where the big runs of salmon occur. This species is sometimes called the king of the salmon because the indigenous people of North America living on the coast had similar beliefs to the herring fishers of Europe. They believed that the whiptail ribbonfish led the salmon on their migrations to spawn at the rivers where they hatched, much as the king of the herrings was once thought to lead the herring shoals in Europe.

OARFISH

CLASS **Osteichthyes**

ORDER **Lampridiformes**

FAMILY **Regalecidae**

GENUS AND SPECIES ***Regalecus glesne***

ALTERNATIVE NAME
King of the herrings

WEIGHT
Up to 600 lb. (270 kg)

LENGTH
Up to 36 ft. (11 m); perhaps longer

DISTINCTIVE FEATURES
Reputedly longest of all fish. Long, slender, compressed body; dorsal fin runs along entire length of body, front rays of which form crest; concave head profile; highly protrusible mouth; tiny spines project from long rays of tail and pelvic fins; metallic silver in color with blotches and wavy markings on body; pink or red fins.

DIET
Small crustaceans such as euphausids; also small fish and squid

BREEDING
Little known. Breeding season: spawning July–December.

LIFE SPAN
Not known

HABITAT
Open seas and oceans at depths of up to 3,280 ft. (1,000 m)

DISTRIBUTION
Worldwide in all tropical and temperate marine waters including Atlantic, Indian and Pacific Oceans

STATUS
Probably fairly common

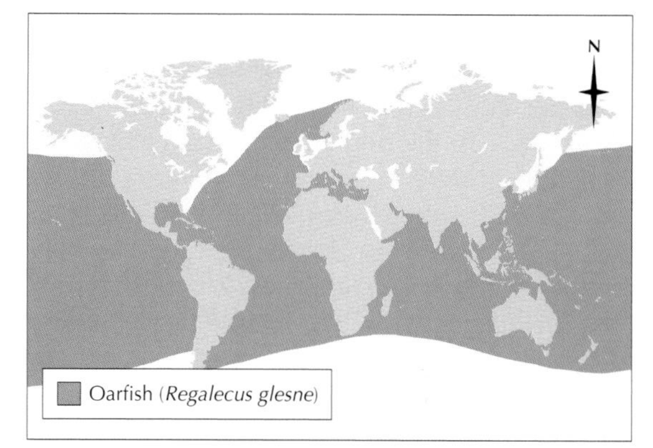

OCEAN

A school of bigeye jacks circles in the sunlit, or epipelagic, zone. Jacks are typical sunlit zone fish, being sleek fast-swimming, visual predators.

While the open ocean can appear to be a featureless, sparsely populated biome to the lay person, it is in reality home to a great variety of life. There is a far wider range of aquatic than terrestrial creatures on Earth, and the majority of them live in the oceans and seas—after all, about 71 percent of the planet is covered in water, mostly seawater.

Humans have scaled all the world's heights and traversed all its landmasses, yet only a small fraction of the oceans have been explored. Most studied are the relatively shallow coastal waters that cover the continental shelves, which are submerged extensions of the major landmasses. Continental shelves vary in width, but extend on average 45 miles (70 km) offshore. These waters are generally no more than 650 feet (200 m) deep. The shelves end at steep continental slopes, which veer sharply down to the much deeper ocean floor. Some of the best-known marine habitats lie in coastal realms, including coral reefs and kelp forests. Habitats of the open ocean—those that lie beyond the continental shelves—are not so well known. Parts of the ocean are as difficult to get to as the moon, with freezing temperatures and immense pressures. However, advances in submersible technology have opened up even the deepest ocean trenches to exploration, albeit of a limited nature. Some surprising discoveries have been made, including ecosystems that, unlike any others on Earth, do not need sunlight.

Layer of light

Although other factors play a major part in ocean ecosystems, including temperature, current, nutrient supply and salinity, the amount of sunlight is the most important. Marine biologists divide the ocean into different layers based on depth and the amount of sunlight. The narrowest but most productive zone is the top layer: the so-called sunlit, or epipelagic, zone. Sunlight is at its strongest in the water column for the first 700 feet (215 m) or so, after which it diminishes rapidly. This depth is generally considered to be the limit of the epipelagic zone.

The vast majority of all marine life lives in this topmost layer. Epipelagic species comprise two types of organisms: plankton and nekton. All the organisms that drift in the ocean currents, too weak or too small to greatly influence their direction, are considered plankton. This includes not only the microscopic phytoplankton and zooplankton but also much larger organisms that mostly drift with the current, including ocean sunfish and jellyfish-like Portuguese men-of-war.

Producers and consumers

The primary harnessers of energy within an ecosystem are called producers. The main producers of the ocean biome are the tiny plantlike organisms that make up the phytoplankton. Like green plants on land, phytoplankton trap the energy of sunlight and convert it into sugar molecules via the process of photosynthesis. Phytoplankton form the base of virtually all aquatic food webs. They are eaten by primary consumers, the tiny animals and animal-like protists known collectively as zooplankton, which in turn are eaten by larger creatures, and so on. Unusually warm waters, or a sudden influx of nutrients, can give rise to fast-growing phytoplankton populations, or blooms. Blooms crash when one or more of the nutrients are used up. Most blooms are beneficial but some can be toxic, and the toxins build up in the food chain. Shellfish and predatory fish might accumulate high levels in their bodies, making them dangerous for people to eat. Such bioaccumulation leads to fatal cases of ciguatera poisoning.

Zooplankton are highly diverse in nature. The term encompasses single-celled protists, near-microscopic crustaceans called copepods and the larvae of many marine creatures, including fish, crabs, lobsters, sea anemones and shrimps. Immature animals are only temporary members of the zooplankton community. Some organisms, such as the amoeba-like eugelenoids, can perform photosynthesis but will also hunt for food if the levels of light in the water drop.

The term nekton covers all those creatures that, unlike plankton, can propel themselves against the current. This includes many varieties of shoaling fish—mackerel, pilchards, anchovies and herring, for example—as well as those animals that feed on them, including barracudas, marlins and dolphins. Fast-swimming visual hunters such as sharks and tuna are perfectly adapted to hunt in the epipelagic zone. Torpedo-shaped bodies slip easily through the water, and powerful tails provide thrust. Baleen whales, including the humpback and the blue whale, are peaceful giants that filter-feed on the plentiful smaller organisms, including tiny, shrimplike krill and fish, that live in these surface waters.

In the wide-open epipelagic zone there are few places for animals to hide, and many miles may have to be covered in the search for food. Some species take refuge in groups: an individual's likelihood of escape is greater if there are others around to tempt or distract a hungry predator. Some predators, in turn, group together to hunt more effectively. A group of dolphins will herd fish into tight-knit clusters before taking turns to swim through and eat their fill. Many sharks, however, are lone predators that prefer not to have to share their kills. Both cooperative and solitary lifestyles emerged in response to the scattered and widespread nature of the ocean predators' food supply.

Sharks, such as this blue shark (front) and mako shark, are among the oceans' top predators.

Jellyfish feed on plankton in the sunlit and twilight zones. The shimmer of the night sea is often caused by the bioluminescence that these animals emit.

Another form of defense is countershading. As sunlight filters through water, any animals present block the rays' passage and appear as distinct silhouettes from below. Fish of the sunlit zone are often lighter on their underside than their topside, though, canceling this effect out.

Twilight zone

Beneath the epipelagic zone lies the twilight, or mesopelagic, zone. Extending down to around 3,300 feet (1,000 m), these waters are only dimly lit by the sun at best. This is where the deep sea can be said to start. Phytoplankton cannot grow without sunlight but zooplankton still occur at these depths, though the species are larger and more luminous. Some take refuge from visual hunters in the mesopelagic zone during the day, rising to the epipelagic zone at night to feed in the darkened waters. They adjust their buoyancy by regulating internal gas bubbles. These daily vertical migrations, which might cover several hundred feet, are mimicked by other animals that feed on them.

The lack of light makes vision less important in this zone, and many species are blind. Farther down or at night, the only light is that produced by certain mesopelagic organisms themselves. Bioluminescence is increasingly common with depth. Light-producing organs can be used to lure prey, warn predators or attract mates. Hatchet fish have light organs on their underside that are thought to be the mesopelagic version of countershading. The organs counterilluminate the fish so that, when seen from below, their lights replicate that filtering gently from above.

Gelatinous predators such as siphonophores, jellyfish and comb jellies are common. Siphonophores, which are colonies of polyps, not single organisms, up to 130 feet (40 m) long have been recorded. Gigantism is, for uncertain reasons, common in deep waters, and several different types of animal grow to a large size.

Deep-sea fish have weak bones because calcium is in short supply and there is no sunlight to produce Vitamin D. Hence, with the high levels of water pressure, deformities are common. Organisms withstand the pressure largely because they are very watery themselves, so external and internal pressures balance out. High levels of water in their tissues make many animals transparent.

Below 1,600 feet (480 m), many organisms are red, purple, brown or black. Since the only light, if any, is the faint blue-green parts of the spectrum that penetrate farthest down, red and dark objects are nonreflective and appear to be black—if they can be made out at all.

Many fish hunt in both the epipelagic and upper mesopelagic zones, as do several species of dolphins. Although dolphins have to surface

A sperm whale takes a breath at the surface before diving to depths of up to 8,000 feet (2,400 m). Sperm whales frequent the depths in search of giant squid, a favorite food.

to breathe, they do not solely rely on sight to hunt. They can also use reflected sound to echolocate prey, and so are not limited to feeding in surface waters. Fish such as sharks have electroreceptors that enable hunting in darker waters. Lantern fish and several species of anglerfish are largely mesopelagic. Fish unique to deeper waters are generally not so well streamlined as their surface-living counterparts. Ambushing prey is more common than swimming in pursuit: food is scarce, and energy needs to be conserved. Large mouths and extendible stomachs are further adaptations that allow deep-sea fish to eat whatever comes their way.

Midnight zone

Below 1,000 feet (300 m) are the pitch-black, near-freezing waters of the midnight, or bathypelagic, zone, which extends to around 4,000 feet (1,200 m). The inhabitants of this zone are widely dispersed, and food scarcity is a great problem. As in the mesopelagic zone, some feed on dead animals, feces and other organic matter that rains down from above. The water is also oxygen poor, and many animals cope by being sluggish and respiring very slowly. Shipworms (a type of bivalve mollusk—a group that includes clams) can live at great depths, where they are able to use the starch glycogen as an oxygen substitute when necessary. Large species of squid and octopus are common, and some of the deepest waters may be home to the world's largest invertebrate: the giant squid. Although giant squid reach lengths of more than 60 feet (18 m), a live specimen has never been seen—only dead remains washed ashore or found partly digested inside a sperm whale's stomach testify to their existence.

Abyssal and hadal depths

Ocean waters that reach depths of greater than 2½ miles (4 km) form the abyssopelagic zone. Again, fish and mollusks like squid and octopus live there, as well as swimming sea cucumbers and the deep-sea relatives of jellyfish. The only places deeper than the abyssopelagic zone are ocean trenches, known as hadal zones. At 36,201 feet (11,034 m) deep, the Mariana Trench in the west Pacific is the deepest ocean trench. However, the difficulty in studying trench life forms leaves many questions unanswered.

Benthic zone

The ocean floor and its immediate environs are called the benthic zone. A sea- or ocean-floor community of any depth can be described as benthic. On average, the ocean floor is from 2½ to 4 miles (4 to 6 km) below the surface. The main inhabitants of this zone are invertebrates such as sea cucumbers, clams, tusk shells, worms, crabs (including giant spider crabs), sea urchins, sea anemones, sponges, brittlestars and sea lilies. Benthic animals feed on each other and on detritus that rains down from above: there are few energy producers on a deep ocean floor. In the Gulf of Mexico, giant isopods search out dead and injured animals to feed on, while golden crabs feed on rancid flesh. Spikes on the spindly legs of deep-sea crabs stop them from sinking in the mud. Rattails, or grenadiers, are common fish on and around ocean floors.

Until the 1970s scientists believed that producers harnessing sunlight were at the base of every food web on Earth. Then, in 1977, below the surface of the Pacific Ocean the crew of a submersible discovered a hydrothermal vent. These underwater geysers form on the midocean ridge, which separates the crusts of Earth's oceanic plates. Cold water seeps into cracks along the ridge and deep into the crust. Heated and under great pressure, it re-emerges at vents. Minerals and chemicals that have dissolved into the water from the rocks form plumes when they mix with the cold seawater. Despite the presence of toxic chemicals and boiling-hot water, many animals were found living around such vents.

It is now known that the producers in these communities are chemosynthetic bacteria. Chemosynthesis involves the breakdown of chemicals—in this case, from the vent water—to create energy molecules. The bacteria live inside giant tubeworms that grow attached to the vent. The worms extract oxygen and hydrogen sulfide from the water, and the bacteria use this for chemosynthesis. In turn, the worm is nourished by a constant supply of bacteria. This type of mutually beneficial relationship between organisms is an example of mutualism. Vent clams have a similar type of symbiotic relationship with bacteria. Spider crabs, shrimps and fish come to graze on the vent organisms.

Other ocean floor ecosystems with similar producers occur where gases and chemicals seep from faults in the oceanic crust. Millions of tubeworms may live in such places. Underwater brine pools are areas of very high salinity so dense that submersibles float on them. Animals that stumble in are killed, but mussels line the edges of the pools. They contain symbiotic bacteria that feed on the methane-rich waters and nourish the mussels. Starfish, eels and other predators visit the mussel beds to feed on their soft flesh. Sometimes, methane or other gases seeping from the crust form hydrate mounds. These crystalline structures form at very low temperatures and high pressure. Ice worms live on and inside the mounds, but scientists have yet to figure out what they feed on.

Threats

Recent research has shown that deep ocean floors have levels of biodiversity that approach those of rain forests. Yet, these irreplaceable habitats are still being used as dumping grounds for radioactive and other types of toxic waste. Deep-sea sediments have been found to contain high levels of pollutants, proving the far-reaching effects of human activity. To continue using the oceans as receptacles for humanity's garbage can only be problematic for their survival.

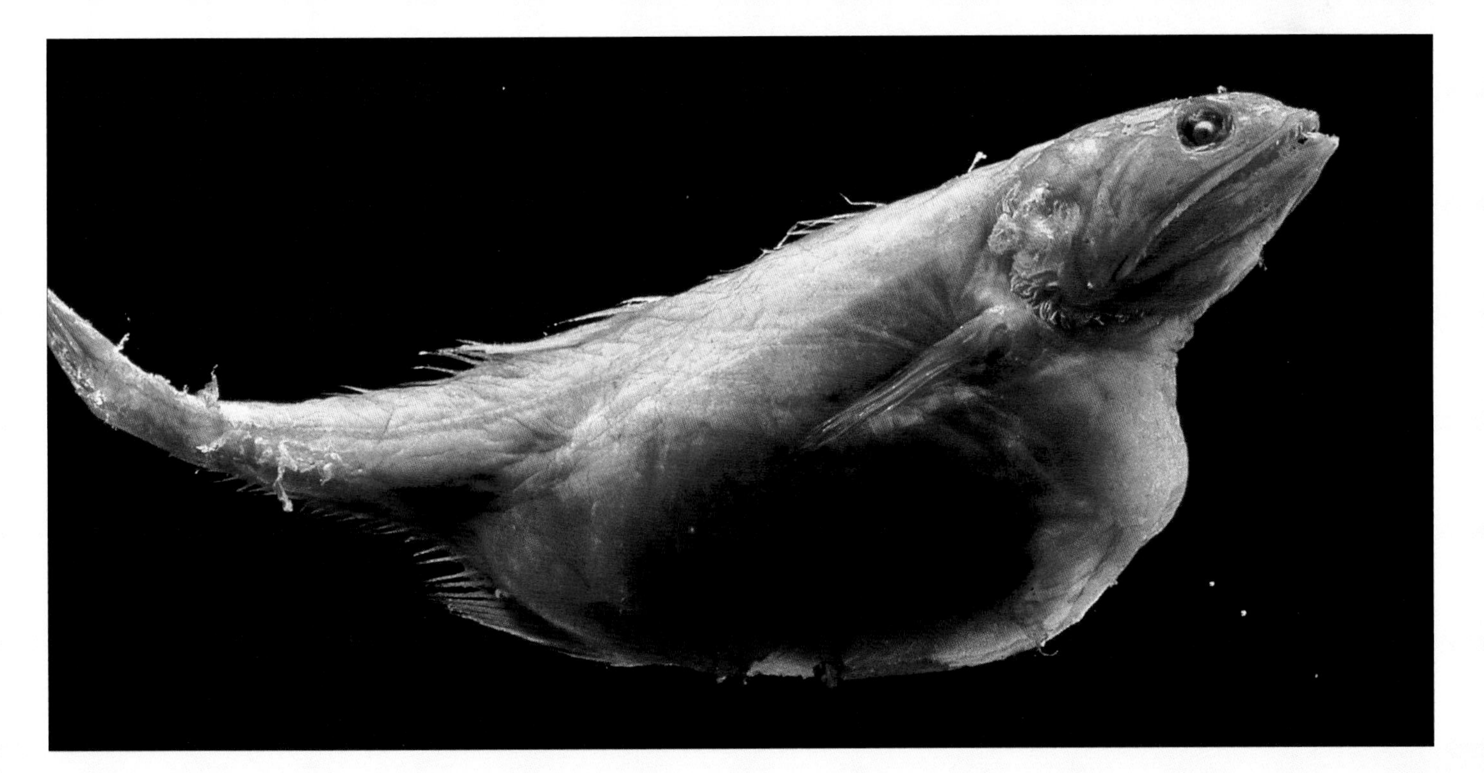

A black swallower,* Chiasmodon niger, *with a full stomach. Species in the family of deep-sea swallower fish, Chiasmodontidae, are distinctive for the way their lower jaw almost disarticulates to allow large prey to be swallowed.

OCEAN SUNFISH

Several ocean sunfish off the coast of California. Sunfish, so called because of their highly distinctive shape, are the heaviest of all bony fish.

THE GENERIC AND SPECIES NAME of the ocean sunfish, *mola*, is Latin for "millstone," a reference to the fish's circular body shape. Sunfish are regarded as the heaviest of all bony fish and are distributed throughout warm and temperate waters worldwide.

The ocean sunfish, *Mola mola*, has a spinal cord that is only about ½ inch (1.3 cm) long, which is shorter than its brain. However, it is the largest of all sunfish, growing to 10¾ feet (3.3 m) and to a maximum weight of about 2¼ tons (2 tonnes). The fish's body is oval and covered with a thick, leathery skin that is bluish, gray, olive brown or nearly black with silvery reflections. The snout projects beyond the small mouth, and the teeth in both upper and lower jaws are joined to form a single sharp-edged beak. The dorsal and anal fins are large and high, and the body ends abruptly in a low tail fin. There are no pelvic fins. The tail of the ocean sunfish is rounded and wavy, but in related species it has a slightly different shape. In the sharptail sunfish, *Masturus lanceolatus*, the tail is drawn out into a point in the middle, and in the slender sunfish, *Ranzania laevis*, it has a rounded margin. The first of these species grows up to 10 feet (3 m) long and weighs about 1 ton (0.9 tonne), while the second seldom exceeds 2 feet (60 cm) in length. A fourth sunfish species, *Mola ramsayi*, is found in the Southern Hemisphere, in the waters around South Africa, Australia, New Zealand and Chile.

Floating at the surface

Ocean sunfish sometimes lie somewhat obliquely at the surface, with the dorsal fin above the surface, as if basking in the sun. They have also been seen well upstream in rivers on a number of occasions, as if they have been carried in on the tide. In October 1960, in Monterey Bay, California, a very heavy death rate was recorded among ocean sunfish close inshore. Skin divers discovered about 100 of the fish in 50 feet (15 m) of water. All of the fish had their fins bitten off and most of them had lost their eyes. On the same day, a little distance from that spot, 20 or more ocean sunfish were seen floating on the surface. These fish had also lost their fins and their eyes had been damaged. The cause of death was never confirmed, but the naturalist Daniel W. Gotshall, who investigated this event, concluded that when sunfish are seen floating at or near the surface, it is likely that they are sick or dying, rather than that they have come to the surface to bask in the sun.

Underwater observations made by the Italian naturalist L. Roghi seem to bear this theory out. He reported that when the sunfish is at rest, it lies stationary in the water with its tail down and its mouth pointing upward. While in this position, it turns a darker color, except for the fins and a large area around the throat. Roghi also reported that as soon as the fish starts to swim, it immediately, and dramatically, changes to a much lighter color. From this evidence, if sunfish came to the surface to bask, it would be logical to assume that they would be light in color and would rest nose up.

Although they have often been sighted at or near the water's surface, especially in calm weather, ocean sunfish may possibly go down to depths of about 1,310 feet (400 m). They are usually seen singly or in pairs, although they may come together in schools of a dozen or more at certain times of the year.

Steering by water jets

The ocean sunfish directs itself by waving its dorsal and anal fins in unison from side to side, in a sculling action, the fins twisting slightly as they wave. The small pectoral fins flap continually but they probably act only as stabilizers. The sunfish uses its tail as a rudder, while it steers with its gills by squirting a strong jet of water out of one gill opening or the other, or out of its mouth. Sunfish do not need to move quickly through the water because their prey is slow-moving. They feed primarily on plankton, jellyfish and other soft-bodied invertebrates, planktonic mollusks, small crustaceans and fish larvae, although they also take a variety other prey small enough to be taken in the beaklike mouth. As well as having a short spinal cord, the sunfish also has a very small brain. It is smaller than either of the fish's two kidneys, which lie just behind it instead of farther back in the body, as is more common among fish.

OCEAN SUNFISH

CLASS **Osteichthyes**

ORDER **Tetraodontiformes**

FAMILY **Molidae**

GENUS AND SPECIES ***Mola mola***

WEIGHT
Up to 2¼ tons (2 tonnes)

LENGTH
Up to 10¾ ft. (3.3 m)

DISTINCTIVE FEATURES
Large, oval body; small eyes; small, beaked mouth; dorsal and anal fins at rear; curved pectoral fins; bluish or gray-brown in color

DIET
Fish larvae, mollusks, zooplankton, jellyfish, crustaceans and brittle stars

BREEDING
Number of eggs: up to 300 million

LIFE SPAN
Not known

HABITAT
Warm and temperate zones of oceans

DISTRIBUTION
Eastern Pacific: British Columbia, Canada, south to Peru and Chile. Eastern Atlantic: Scandinavia south to South Africa; occasionally also western Baltic and Mediterranean. Western Atlantic: Newfoundland, Canada, south to northern South America.

STATUS
Not threatened

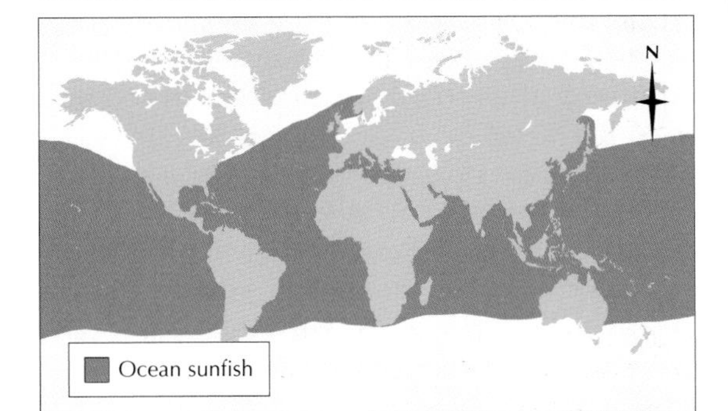

Prolific egg-laying

From dissection of captured sunfish scientists estimate that the female's ovary may contain 300 million eggs, making the sunfish the most fecund fish species. The larvae are about 3 millimeters long and initially have a much more conventional fish shape than the adult sunfish, with large pectoral fins. However, the dorsal and anal fins soon begin to grow and the body becomes covered with spines. At this point in their development the juveniles resemble pufferfish or boxfish, to which sunfish are related. The spiny coat is then gradually shed, until only five long spines are left. These shorten until they are lost completely, and the bulky, disc-shaped body begins to take shape when the young fish is about ½ inch (1.3 cm) long. At this stage the parrotlike beak develops also, and the fish gradually begins to assume the typical characteristics of an adult sunfish.

Ocean sunfish often drift at the water's surface, sometimes with their dorsal fins showing (below). Some naturalists regard this as abnormal behavior, and suggest that it is a sign of sickness.

OCELOT

The ocelot is a medium-sized, long-legged cat found from Texas southward into most of South America. A solitary animal, it hunts mainly at night for small mammals, reptiles and birds.

ONE OF THE MOST ATTRACTIVE members of the cat family, the ocelot's short-haired coat is sandy, grayish yellow or deep, warm brown in color, blotched with large brown, black-bordered rosettes and spots. It has two black cheek spots around the head and neck. The ocelot has very long legs and is medium-sized, being 1⅘–3¼ feet (0.5–1 m) in head and body length, with an 8–10-inch (20–25-cm) tail. It weighs up to 26½ pounds (12 kg). The name ocelot is from a Mexican word *tlalocelotl*, meaning field tiger. The pattern of the species' coat provides good camouflage in the cat's native forests and so helps it considerably when hunting. There are rare melanistic (all-black) forms.

Closely related to the ocelot is the margay cat, *Leopardus wiedi*. It is considerably smaller, about the same size as a domestic cat, and is bright creamy yellow in color, spotted with jet black. The margay cat weighs up 7 pounds (3.2 kg) and has a very long tail.

The ocelot is found in the southwestern United States, in Trinidad, throughout Mexico and Central America and in South America apart from Chile. The margay is found over most of this range but is restricted to the forested areas. It is less numerous than the ocelot.

Hunting in the thickets

A forest-loving animal, the ocelot generally keeps to dense cover. Away from the forest it might be found in thorn scrub or pastures. Although it can climb well, the ocelot normally hunts on the forest floor, making good use of its acute hearing and sight. Unlike most cats, it swims well. It is entirely nocturnal, hiding in tree holes during the day. The ocelot confines its hunting to a more or less fixed territory, which it defends against its own kind, although ocelots are sometimes seen hunting in pairs. They mark their territory with urine and feces.

Efficient hunters

The ocelot preys on almost any animal it can overpower. These include small mammals, such as rats, mice, armadillos, lesser anteaters and agoutis. It will even take monkeys and young deer. Birds, reptiles and frogs are also eaten, as are spawning fish because they are easy to catch. The ocelot can be a threat to domestic stock and poultry, sometimes taking lambs and pigs.

The margay hunts mainly small, arboreal (tree-dwelling) mammals and birds, but it also eats some fallen fruit. It has rotational hind feet, adapted for climbing trees.

OCELOT

CLASS **Mammalia**

ORDER **Carnivora**

FAMILY **Felidae**

GENUS AND SPECIES ***Leopardus pardalis***

WEIGHT
Male: 22–26½ lb. (10–12 kg); female: 19½–20¾ lb. (8.8–9.4 kg)

LENGTH
Head and body: 1⅗–3¼ ft. (0.5–1 m); shoulder height: 1⅔ ft. (0.5 m); tail: 8–10 in. (20–25 cm)

DISTINCTIVE FEATURES
Medium-sized cat with very long legs. Sandy yellow, grayish yellow or warm brown coat, marked with brown or black rosettes or spots; 2 black cheek spots; occasional melanistic (all-black) forms.

DIET
Small mammals such as mice, rats, agoutis, armadillos, lesser anteaters and young deer; also birds, reptiles, frogs and spawning fish

BREEDING
Age at first breeding: 30 months (male), 18–22 months (female); breeding season: all year, with peak in fall; number of young: usually 1; gestation period: 79–85 days; breeding interval: probably 2 years

LIFE SPAN
Up 10 years

HABITAT
Mainly forests; also thorn scrub, forest edges and pastures

DISTRIBUTION
Southwestern U.S. south into Central and South America, including island of Trinidad; absent from Chile

STATUS
Fairly common in most of range

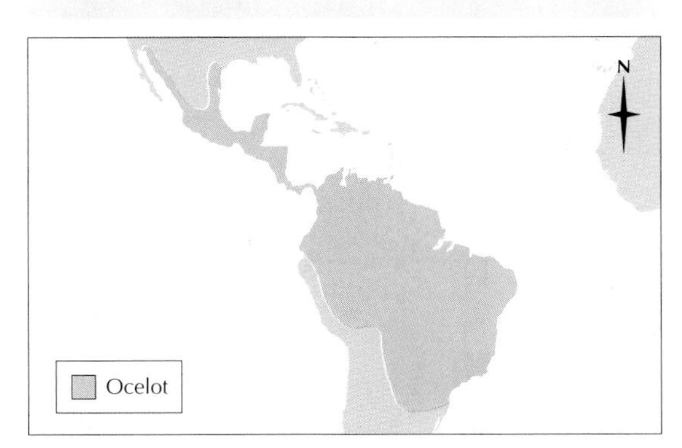

Ocelots were once under threat from the fur trade, but legal protection and hunting bans are now in force.

Small litters

The breeding season for the ocelot is year round, but with peaks in the fall. After a gestation period of 79–85 days, a single kitten is born. Occasionally there are there are two in a litter, and very rarely three. The kitten is born well-furred but blind, in a nest made of grass or other soft material in a hollow log, among rocks or under a bush. The female ocelot probably has only one litter every 2 years.

Dangerous beauty

Although the ocelot's coat makes it inconspicuous where it lives, it has the disadvantage that it makes desirable fur coats. The fur is expensive as no two ocelots have identical markings, and fur dealers have great difficulty in matching the pelts. It is now illegal to hunt ocelots or trade their pelts in the United States and in most other countries where they are found. Although declining in numbers they remain quite common, with a wild population of 1.5 to 3 million. Some subspecies are endangered, however, and margays are becoming rare in some parts of their range.

Two hunting strategies

Most members of the cat family, including the ocelot and the domestic cat, stalk their prey. When it first sees a mouse or small mammal at some distance, a cat will crouch down and then slink quickly toward it with its body flat to the ground. When it gets near, the cat will pause and watch. It then stalks its victim, moving forward slowly and cautiously and using any available cover until it launches its final attack. A single leap is most likely to be successful, but this maneuver may be preceded by a short run with the body flat to the ground. The prey is killed with a single bite at the nape of the neck.

The hunting behavior just described is an adaptation for catching small rodents rather than birds. The ocelot, in contrast with most other members of the cat family, is also highly successful at killing birds. This is because it makes a direct attack, rather than an ambush, the instant it sees a suitable bird within range.

Once popular as pets

Owing to their attractive appearance, the ocelot and the margay used to be frequently taken when very young and reared by hand as domestic pets, especially in the United States. The initial attractiveness of the margay as a pet springs to some extent from the large size of its eyes, which are unusually large even for a cat. Both the margay and the ocelot can become unpredictable, however, and are often dangerous when they grow to adult size. Moreover, the average home cannot provide a suitable enviroment for these species.

An ocelot in its daytime resting place on the forest floor. The spotted coats of ocelots mimic the effect of dappled sunlight on leaves.

OCTOPUS

The octopus has eight arms, joined at their bases by a web and surrounding a beaked mouth. These arms can be regrown if they are lost in a fight. Octopuses differ most obviously from squid and cuttlefish, the other well-known members of the class Cephalopoda, in lacking an extra pair of long tentacles. Their suckers, which run along the arms, are not strengthened by the horny rings seen in the suckers of squid. Finally, octopuses have no trace of an internal shell, and their bodies are short and rounded instead of streamlined. The two large eyes are located at the top of the body sac.

The 150 species of octopuses are distributed throughout the seas of the world but are especially numerous in warm waters. The smallest species is *Octopus arborescens,* which is less than 2 inches (5 cm) across. The largest is the Pacific octopus, *O. hongkongensis,* which reaches 32 feet (10 m) across the arms, although its thimble-shaped body is only 18 inches (46 cm) long. Another giant is *O. apollyon,* of the North Pacific, which reaches 28 feet (8.5 m) across. The order Octopoda includes other species that are markedly different in form. One is the argonaut of the genus *Argonauta,* while another is the blind deep-sea cirrothauma of the genus *Cirrothauma,* which is found in the North Atlantic and has two large fins on its body. The web between the cirrothauma's arms reaches almost to their tips and the species swims by opening and closing this umbrella-like structure. Besides the suckers on the undersides of its arms, it has rows of filaments that are used for catching food particles. Its body is transparent and has the texture of a jellyfish.

The common octopus, *O. vulgaris,* lives off the coasts of tropical and subtropical Africa, the Mediterranean and Atlantic America, and also reaches the southern coasts of Britain. It may exceptionally reach a span of 10 feet (3 m) but is usually much smaller. The lesser octopus or curly octopus, *Eledone cirrhosa,* ranges from Norway south to the Mediterranean. It is rarely more than 2½ feet (75 cm) across the widest span of its arms and is identifiable by the single row of suckers on its arms in contrast to the double row in the common octopus.

Octopuses (*Octopus cyanea*, above) are among the most intelligent of all invertebrates. They have well-developed eyes and can sense different textures with their arms.

Masters of disguise

The common octopus lives among rocks in shallow water, spending much of the time in a hole in the rocks or in a lair built of stones. When outside, it creeps about on its arms most of the time, using its suckers to grip, although it can also swim. It usually swims by blowing water out through its siphon, its arms trailing behind to give a streamlined shape. As in cuttlefish and squid, this water is blown out from the mantle cavity, which houses the gills and the openings of the kidneys, rectum, reproductive organs and ink sac. Like cuttlefish and squid, the octopus can send out a cloud of ink from its rectum to baffle pursuers.

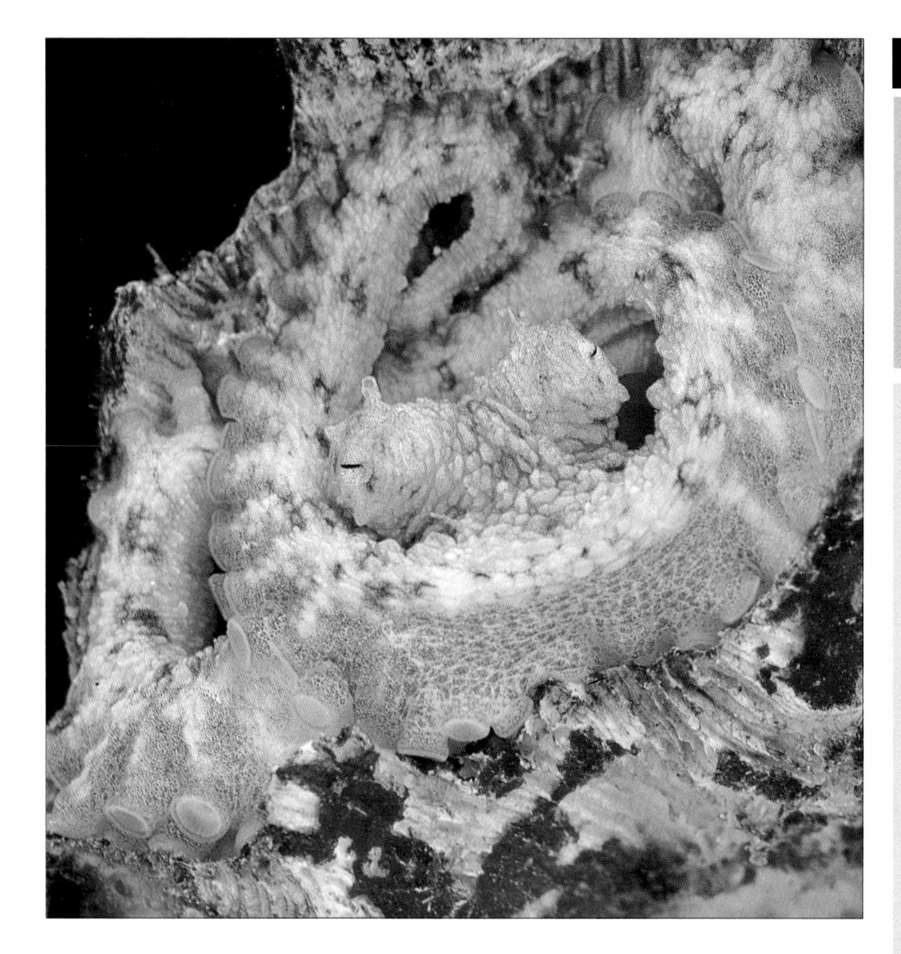

Like many other cephalopods, the common octopus is able to vary its color, size and texture in order to blend in with its background.

Chameleon-like color changes

There is no evidence that an octopus reacts to sound. However, the arms are very sensitive to touch and taste, and the eyes are well developed. The importance of vision is reflected in the octopus's ability to change color. This feat is achieved using two kinds of chromatophores, or pigment cells, in the skin, which vary in color according to the degree to which they are expanded or contracted. One kind of cell varies from black to reddish brown and the other from red to pale orange yellow. Beneath these chromatophores is a layer of small bodies, known as iridocytes, that reflect white light or give a blue or green color by refraction.

However, variation in the octopus's appearance is not only brought about by a change in color patterns but is also influenced by alterations in posture and general texture. The arms may be extended, tucked underneath or curled stiffly back over the body as armor, and the suckers may be out of sight or protruded to give the arms a wavy outline. By suitable adjustments in color, posture and texture an octopus can merge completely with its background and so become extremely difficult to see.

Octopuses are also able to effect a conspicuous display that often gives them away to fishers searching for them. This is known as a dymantic display and occurs when octopuses are frightened by large objects. The animal flattens out, coiling its arms in beside the body and extending the web between them. The body grows pale but dark rings develop around the eyes and the edge of the web also becomes dark. The purpose of this display is probably to deter predators, at least long enough for the octopus to change color, blow ink and swim away.

With their large brains and adaptable behavior, octopuses have been the object of a number of revealing studies on learning and

COMMON OCTOPUS

PHYLUM **Mollusca**

CLASS **Cephalopoda**

ORDER **Octopoda**

FAMILY **Octopodidae**

GENUS AND SPECIES ***Octopus vulgaris***

LENGTH
Maximum arm span: 10 ft. (3 m)

DISTINCTIVE FEATURES
Eight thick arms, joined at bases by weblike membrane; rows of suckers along arms; round, warty body without internal shell; beaklike mouth; 2 large eyes on top of body

DIET
Crabs, lobsters, mollusks, snails and fish

BREEDING
Number of eggs: about 150,000; hatching period: 4–6 weeks; breeding interval: 1 year

LIFE SPAN
About 2 years

HABITAT
Rocky seabeds in shallow waters

DISTRIBUTION
Mediterranean, Red Sea and coastal waters of the Atlantic Ocean and Caribbean

STATUS
Common

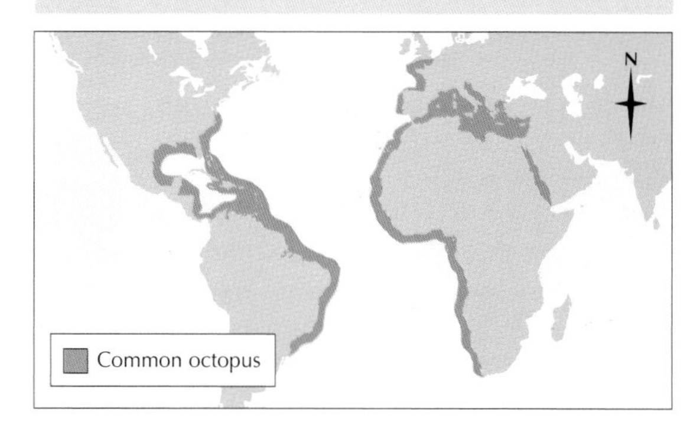

brain function in lower animals. They are able to distinguish texture with their sensitive arms, can be taught to recognize different shapes and are able to remember experiences for several weeks. In captivity they rapidly settle down and become accustomed to their new environment.

Octopus attack

An octopus usually attacks only moving objects. It glides smoothly to within a few inches of its prey, collects its arms together and then jumps forward at the victim with a sudden backward spurt from its water jet. Small prey, mainly fish and crustaceans, are trapped underneath the expanded web between the arms and then seized with the parrotlike horny beak around the octopus's mouth. At the same time the octopus emits a poison that paralyzes the prey, before it commences to eat. It often takes the stunned prey back to its lair before eating, throwing any shells outside the lair's entrance. An average-sized octopus eats perhaps two dozen small crabs in a day. It also drills holes in a wide range of mollusks, ranging from chitons to bivalves.

There are many reports of divers being seized and held by octopuses, and there is little doubt this does happen occasionally, particularly in warm seas, though such incidents are probably rare. It seems, however, that these are not deliberate attacks but more a matter of the octopus investigating a moving object, and if the divers keep still, the octopuses usually "feel" them for a short while and then let them go. Although most octopuses are harmless to humans, the blue-ringed octopus, *Hapalochlaena maculosa*, which grows to only 4 inches (10 cm) in length, is capable of emitting a deadly poison that can kill a human in 15 minutes.

Restricted courtship

Males and females change color to attract each other's attention during courtship. There is almost no other premating display, although the male may expose certain particularly large suckers near the base of the second pair of arms. The only contact he has with the female is through a single arm, which he extends to caress her. This arm is always the third arm on the right side, which is specially modified for the purpose and has a spoon-shaped tip. The tip is placed in the female's gill cavity and the male's sperms are deposited near the opening of her oviduct in elaborate packages known as spermatophores.

A female may lay up to 150,000 eggs in a period of about week, each in an oval capsule slightly smaller than a grain of rice. They are attached by short stalks to long strands that festoon the roof of the mother's lair. The mother broods over the eggs for several weeks, often cleaning them with her arms or blowing water over them with her funnel to keep them free of algae and parasites. During this time, she eats little. Indeed, she may fast completely for weeks, or for as long as 4 months in one species.

Female dies after her young hatch

The short-armed young hatch at about 3 millimeters long. They drift around for a while before they start their own life on the bottom, by which time they may be ½ inch (1.3 cm) long and several weeks old. The female usually dies shortly after the young have hatched. The young feed on tiny organisms until they have developed to a size at which they are able to take larger prey.

The common octopus rarely if ever breeds on British coasts, although year after year larvae migrate across the English Channel from Brittany on the northern coast of France. Sometimes after a mild winter the numbers of octopuses may reach plague proportions, to the detriment of local crab and lobster populations.

An octopus uses its many suckers to hold onto slippery prey and to maintain its grip as it moves over the seabed. Pictured is a small specimen of the common octopus.

OIL BEETLE

A female oil beetle, Meloë proscarabaeus, with her newly laid batches of orange eggs. Although there are probably 3,000 to 4,000 eggs in each batch, few of the larvae will survive to adulthood.

These small beetles get their name from their habit of exuding an oily fluid from the joints of their legs when alarmed. Oil beetles belong to the same family as the blister beetle and have a similarly complicated life cycle. However, they are clumsy insects and do not have the bright colors of the blister beetle. They were once described by the famous French entomologist Jean Henri Fabre as "uncouth beetles... their wing cases yawning over their back like the tails of a fat man's coat that is far too tight for the wearer." This describes the oil beetles' very short, soft elytra, or wing cases, that do not cover the abdomen and overlap in a most unusual way for a beetle. The wing cases are, in fact, functionless as these beetles have no wings.

A common European oil beetle, *Meloë proscarabaeus,* is 1 inch (2.5 cm) long and bluish black in color. In the deserts of the southwestern United States and Mexico there are some more unusual oil beetles, about 1–1½ inches (2.5–3.8 cm) long, with very hard bodies and wing cases. The wing cases are fused together and are larger than usual, reaching over the abdomen to form an air chamber. The air in this chamber acts as an insulating barrier against the sun's heat, while the hemispherical shape of these beetles presents a relatively small surface area for the absorption of heat. The hard, thick body armor also prevents evaporation of the body fluids into the dry desert air.

Hitchhiking thieves

Relatively low numbers of oil beetle larvae survive to adulthood because the larval life is so complicated. To survive, the larva must manage to hitch a ride, usually on the body of a solitary bee, and then drop into a cell of its honeycomb.

The stout antennae of the adult males of some species resemble the "forceps" of an earwig and are used to clasp the female during mating. As the thousands of eggs develop within her abdomen, the female becomes extremely swollen. She eventually lays several batches of 3,000 to 4,000 eggs in cracks or holes in the ground. They hatch after 3–6 weeks, and the thousands of tiny larvae emerge to swarm up the stems of surrounding plants.

The hatchlings look like lice, with long, narrow bodies, and when they were first discovered hanging on to the bodies of bees, they were called bee lice. They are very active and scramble up plants to sit in the flowers until a solitary bee visits. When a bee alights, the larvae grip it with their jaws

OIL BEETLES

PHYLUM **Arthropoda**

CLASS **Insecta**

ORDER **Coleoptera**

FAMILY **Meloidae**

GENUS AND SPECIES ***Meloë proscarabaeus; Apulus muralis; Epicauta velata*; others**

LENGTH
Average ⅖–1⅕ in. (1–3 cm)

DISTINCTIVE FEATURES
Clumsy-looking beetle. Short, soft elytra (wing cases) that gape open at rear; dull coloration; wingless.

DIET
Adult: variety of plants. Larva: eggs of insects such as bees; pollen and nectar.

BREEDING
Complex breeding strategy, with 2 different larval stages. Breeding season: eggs laid in spring; number of eggs: several batches of 3,000 to 4,000; hatching period: 3–6 weeks.

LIFE SPAN
Up to about 1 year

HABITAT
Anywhere where host (usually a solitary bee) is found

DISTRIBUTION
Virtually worldwide

STATUS
Common

An oil beetle,* Meloë cicatricosus, *in Slovakia. Oil beetles are unusual because their elytra (wing cases) are very short, soft and functionless.

and are carried away. However, the oil beetle larvae show no discrimination and will grasp any hairy insect. Consequently, they are often borne away on beetles, flies, butterflies and honeybees and will not survive. However, if they catch the right species of solitary bee, they are eventually carried to the bee's nest. Here the larvae drop off and enter a cell, where they devour the solitary bee's single egg.

Transform into inactive grubs

The rest of the larval life is spent in this cell, feeding on the nectar and pollen intended for the growing bee. The louselike larva undergoes a radical transformation in this time and becomes a soft, inactive grub with short legs. All it does is feed. The spiracles, or breathing holes, are placed high on the back to be clear of the sticky fluid on which the larva is floating. While in the bee's nest, the larva sheds its skin several times. At the third molt it is known as the pseudopupa and has only minute legs. Two molts later it becomes the true pupa, which resembles the adult in form. The adult oil beetles emerge from the bee's nest the following spring and the complicated cycle starts again.

When adult, these beetles feed on plants and are sometimes pests of crops such as potatoes and tomatoes. This very complicated cycle of events with two very different kinds of larvae, one active and one passive, is known as hypermetamorphosis. In other words, the oil beetle undergoes more changes in form than the usual larva–pupa–adult stages of other beetles.

In the United States some oil beetle species have a simpler life history. The louselike larvae simply run around until they find a mass of locust eggs, which they settle in and devour.

Caustic blood

An oil beetle crawling sluggishly through the grass would seem to be utterly defenseless, but when handled it exudes an oily, caustic fluid containing cantharidin, which has been known to raise skin blisters. This fluid is the oil beetle's blood. The beetle's reaction to disturbance is to compress its abdomen and raise its blood pressure so much that the thin skin of the joints is ruptured and blood squirts out. Once the pressure is released, the blood clots quickly. This behavior is known as reflex bleeding and is practiced by several insects, including ladybugs (genus *Coccinella*) and certain grasshoppers.

OILBIRD

An oilbird squats on its nest on the Caribbean island of Trinidad. Here the species is known as the diablotin, or little devil, perhaps because of the blood-curdling cries that oilbirds make as they leave their caves.

Also called the guácharo or diablotin, the oilbird is classified in its own suborder (Steatornithes) within the order Caprimulgiformes, the order that also contains the nightjars. The oilbird resembles a cross between a nightjar and a hawk. It has an overall length of 15¾–19⅓ inches (40–49 cm), and its long, narrow wings span 3 feet (90 cm). The face, with its discs around the eyes, is very owl-like; the bill is curved like a hawk's and is surrounded by stout bristles at the base. The feet are very weak, and the oilbird rarely perches but clings to the sides of rocks like a swift. The plumage is brown, barred with black and spotted with white. The oilbird lives in Panama and northern South America as far south as Bolivia and on the Caribbean island of Trinidad.

Ghostly cries

There used to be a belief among the people of Trinidad that oilbirds contained the souls of criminals and other miscreants. This belief arose from the eerie, almost human cries of the oilbirds as they emerge from caves in the evenings. Oilbirds roost and nest in caves either in mountain country or on the coast. The roosts are sometimes as much as ½ mile (800 m) back from the cave entrance, where it is pitch dark. The oilbirds roost on rock ledges during the day and emerge with grating and piercing shrieks and screams in the evening to feed. The calls become softer once the birds are outside the cave.

The size of its eyes shows that the oilbird has good vision; undoubtedly it relies on sight while out foraging, as do nightjars and owls. There is not a glimmer of light in the caves, yet an oilbird flies to and from its nest and roost unerringly. It is now known that the oilbird navigates by echolocation in much the same way as do bats, except that the sounds it emits are within the range of human hearing. When oilbirds are flying in their caves, a continuous metallic clicking like that of a typewriter can be heard through the chorus of wails and screams.

Spicy food

Oilbirds are the only nocturnal birds that feed on fruit. They may travel up to 90 miles (150 km) in search of fruit, which they pluck while hovering among the foliage and carry back to digest during the day. After digesting the flesh of the fruit, the birds regurgitate the seeds. By collecting the seeds that accumulated under the nesting ledges, ornithologist David Snow was able to determine the kinds of fruits that the Trinidad oilbirds eat. The seeds were mainly from three kinds of plants: palms, Caribbean "laurels" and incense. The latter two kinds are aromatic and strong-smelling and, as it is known that the part of the brain concerned with smell is well developed in oilbirds, it may be that they find at least some of their food by smell.

Oilbird nests are mainly a paste of regurgitated fruit, together with seeds and droppings. They are built on a ledge on a cave wall, and as they are used for several years they become quite bulky. The 1 to 4 eggs in a clutch are laid at fairly long intervals and are incubated for 32–35 days. The timing of the egg-laying depends on location. Most eggs are laid in April and early May in Trinidad, for example, but in May and June in Venezuela. The chicks hatch with a thin covering of down, although they grow a thicker coat within 30 days. They build up a great store of fat and may be 1½ times as large as an adult. The chicks lose weight when their feathers form.

Oilbird chicks stay on the nest for 88–125 days, an extremely long time. They probably do this because they are fed on nothing but fruit

OILBIRD

CLASS **Aves**

ORDER **Caprimulgiformes**

SUBORDER **Steatornithes**

FAMILY **Steatornithidae**

GENUS AND SPECIES ***Steatornis caripensis***

ALTERNATIVE NAMES
Guácharo; diablotin

WEIGHT
12⅓–17 oz. (350–485 g)

LENGTH
Head to tail: 15¾–19⅓ in. (40–49 cm)

DISTINCTIVE FEATURES
Owl-like head, with discs around eyes; long, hooked bill; long wings; graduated tail; short, weak legs; chestnut brown all over, with numerous white spots

DIET
Fruit, mainly of palms

BREEDING
Age at first breeding: not known; breeding season: mainly April–June; number of eggs: 1 to 3 (Venezuela), 2 to 4 (Trinidad); incubation period: 32–35 days; fledging period: 88–125 days; breeding interval: about 1 year

LIFE SPAN
Up to 12 years or more

HABITAT
Tropical forests; roosts and breeds in caves, usually in mountainous areas but also in sea caves on Trinidad

DISTRIBUTION
Panama and Colombia east to Venezuela, Trinidad and northern Brazil, south to Ecuador, Peru and Bolivia

STATUS
Scarce but not threatened

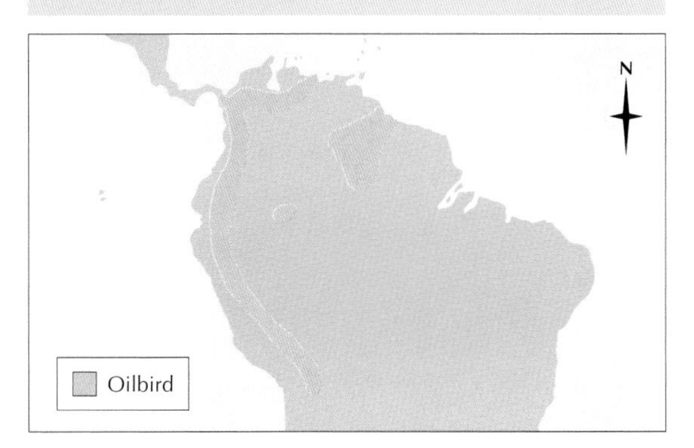

Oilbirds congregate at one of their nesting caves on Trinidad. Oilbird chicks are exploited by local people as a source of cooking oil.

flesh, which contains very little protein. Most fruit-eating birds catch insects to provide their young with protein, but the young oilbirds have to be content with a long nestling period and growing fat from the oily fruit they eat.

In general birds raise their chicks as fast as possible because they are vulnerable in the nest. Although the young oilbirds have no natural enemies in the depths of the caves, people have found a use for them. Their fat can be made into a fine oil that is excellent for cooking purposes and keeps fresh for a year. South American Indians regularly harvest the young oilbirds, knocking them out of their nests with poles.

Clicks in the dark

Donald Griffin, who is well known for his studies of echolocation in bats, investigated the ability of oilbirds to find their way in pitch darkness. He visited an oilbird cave in Venezuela and by exposing photographic film showed that the oilbirds were able to fly where there was absolutely no light. With an oscilloscope he demonstrated that the birds emitted steady trains of clicks at frequencies between 6,000 and 10,000 hertz (cycles per second), well within the range of human hearing. Griffin was even able to hear the echoes rebounding from the cave walls. To show that the oilbirds were navigating by echolocation, Griffin released some in a darkened room with cotton wool in their ears. They were helpless until the cotton wool was removed. Oilbirds' echolocation is not as sensitive as that of bats, which can detect flying insects, but it is certainly sufficient to enable these birds to fly through caves and locate their nests.

OKAPI

One of the last big mammals to be discovered in Africa, the okapi is the giraffe's only living relative. Its neck is not as long as the giraffe's, but the general shape of its body is much the same. It was, however, some time before the relationship between these animals was realized, and at first scientists thought the okapi was an ancestor of the horse.

The okapi is a deep, velvety purple black, becoming lighter with age. The face is whitish except for a dark gray crown and forehead and a blackish muzzle. The lower legs are white with black bands around the fetlocks and foreknees and a black line down the front of the forelegs. On the upper part of the forelimbs are three or four thick, irregular white stripes, and on the upper part of the hind limbs and up onto the haunches are 11 to 19 equally irregular white stripes. The skin is about ¼ inch (6 mm) thick, like a giraffe's, and is rubbery. The male has a pair of short horns on the forehead that develop when the animal is about 2 years old; the female has only little knobs. The horns are covered with skin and tipped with hair in the young.

An okapi rests in the shadows of its forest home. Besides people, leopards are the only major threat to a full-grown okapi.

Inhabits remote forests

The okapi lives in clearings near rivers in the rain forests of the Democratic Republic of Congo (formerly Zaire). Its range covers an area 140 miles (225 km) from north to south and 625 miles (1,000 km) from east to west; approximately from the Semliki River in the east to the Zaire–Ubangui junction in the west. Little is known of the okapi's behavior in the wild except that it moves over well-defined paths. It is thought to be solitary, each animal possibly holding a territory. Normally placid, the okapi can suddenly become aroused, butting and kicking at an aggressor, a frequent occurrence when an okapi is captured.

The okapi eats mainly young shoots and buds, as well as some grasses, leaves, fruits and ferns. The animal is a browser, plucking its food with the tongue, which is long and mobile, like a giraffe's. At the same time the long neck is very supple, and the animal can turn its head to lick every part of its body.

Gestation puzzle

At breeding time, a pair of okapis stays together for about 2–3 weeks, the female being in season for 40–50 days. However, according to a former biologist at the Antwerp Zoo in Belgium, females continue to come into heat while pregnant, so for some time it was difficult to be sure of the gestation period, and all that could be said was that it was 9–15 months. Later observations at Bristol Zoo showed it to be about 450 days (about 15 months). This gestation is about the same length as that of the giraffe, which is 420–468 days—much longer than that of any other ruminant.

Hard to breed in captivity

Until the 1940s about 20 okapis had been taken to zoos. Of these, only two—one in Antwerp, 1927–1942, and one in London, 1937–1950—survived more than a year or two. In 1948 the Antwerp Zoological Society started to organize the capture of okapis for zoos, and set up a special center at Epulu, from which breeding pairs were to be sent all over the world.

The first zoo-born okapi was born in Antwerp in September 1954, but it survived only a day; several others born in various zoos did not live long either. Then in 1958 the center at Epulu

OKAPI

CLASS Mammalia

ORDER Artiodactyla

FAMILY Giraffidae

GENUS AND SPECIES *Okapia johnstoni*

WEIGHT
440–660 lb. (200–300 kg)

LENGTH
Head and body: 6¼–7¼ ft. (1.9–2.2 m); shoulder height: 5–6 ft. (1.5–1.8 m); tail: 12–16½ in. (30–42 cm)

DISTINCTIVE FEATURES
"Front-heavy" appearance, with body much higher at shoulder than at rump; giraffe-like head and legs; purple-black coat with black and white stripes across upper legs and buttocks; tail tuft; short horns (male only)

DIET
Forest vegetation; also fruits and seeds

BREEDING
Age at first breeding: 2 years; breeding season: all year, births peak in rainy season; number of young: 1; gestation period: about 450 days; breeding interval: 1–2 years

LIFE SPAN
Up to 33 years in captivity

HABITAT
Rain forests, often in clearings near rivers

DISTRIBUTION
Parts of Congo Basin

STATUS
At low risk; population: 10,000 to 20,000

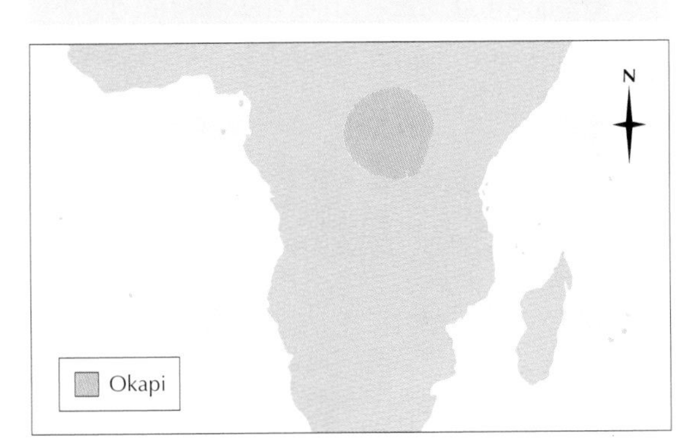

and the Paris Zoo each succeeded in breeding an okapi; both animals lived for many years. At the beginning of 1969 there were 51 okapis in zoos, of which 29 had been born and bred in captivity. In 2000 there were 43 okapis living in captivity throughout the world.

Okapis are believed to be solitary animals, males and females meeting only in the breeding season.

A new discovery

Rumors of the okapi's existence reached Europe before 1900. Explorer and journalist H. M. Stanley wrote in his book *In Darkest Africa* (1890), "The Wambutti know a donkey and call it 'atti.' They say that they sometimes catch them in pits." Harry Johnston, before going to Uganda as governor, spoke to Stanley in London about the "atti" of the Wambutti. Once in Africa, Johnston met Archdeacon Lloyd of the Boga Mission, the only European to have seen an okapi, and he confirmed what Stanley had said.

Johnston continued to pick up tales of the creature and learned the correct spelling of its name, "o'api" (the ' being pronounced as a guttural *k*). Because the Wambutti told him that the animal had more than one hoof, Johnston decided that it must be a survivor of *Hipparion*, the three-toed extinct horse. After trying for a while to get a complete specimen, he purchased two strips of skin from the tribesmen and sent them to P. I. Sclater, secretary of the Zoological Society of London, as skin from a new species of zebra living in the Zaire forest.

In February 1901, Sclater gave the animal the name *Equus johnstoni*. However, shortly afterward, a Belgian officer obtained two skulls and a complete skin, which showed the animal to be cloven-hoofed, not odd-toed and horselike. Therefore it belonged to the giraffe family. In June 1901, Ray Lankester gave it a new generic name, *Okapia*. The great variability in the stripe pattern of okapis led to several supposed new species being described in the early 20th century. Nevertheless *Okapia johnstoni* remains the only recognized species of okapi.

OLM

THE ONLY CAVE-DWELLING AMPHIBIAN in Europe, the olm belongs to the Proteidae, the same family as the mudpuppy, *Necturus maculosus,* of North America. The olm (the *l* is pronounced) lives in underground rivers and pools and, like many other cave-dwelling animals, it is blind and lacks pigment in the skin. It grows to a length of about 12 inches (30 cm), and the eel-like body has a laterally flattened tail. The head is broad with a blunt snout, and the legs are very short, with three toes on the front legs and two on the hind legs. The eyes are minute and buried in the skin; they cannot be seen in the adult. The body is a translucent white, becoming pinker when the olm is active and blood is flowing through the skin. The three pairs of feathery gills are red.

The olm is restricted to limestone caverns in southeastern Europe. It is found in parts of the former Yugoslavia, such as Dalmatia, Bosnia, Herzegovina and Croatia, and in a small area of Italy. It is most common in caverns that form part of the underground course of the Pivka River in Slovenia.

Cave dwellers

The first record of the olm is in a book by the 17th-century Slovene scholar Janez Vajkard Valvasor, who recorded that some peasants ascribed the periodic flooding of the Bella River to a dragon that lived inside a mountain and opened sluice gates when its hideaway was threatened with floods. In the floodwaters the villagers often found lizardlike animals that they took to be the dragon's babies. These animals were olms, and it is only during floods that they are found outside their underground caverns.

Inside the caverns, olms live in still pools or in underground rivers, where they spend most of their time concealed under boulders. They can, however, swim with great agility and are very difficult to catch. Like newts, olms absorb oxygen through their skin, but they also rise to the surface to gulp air. Olms feed on small aquatic animals such as crustaceans and mollusks. They are known to be very sensitive to vibrations, so they probably detect their prey by the movements it makes in the water.

Losing color

It is usually impossible to distinguish the sexes of olms, but the females become plump and pink in the breeding season. Courtship and mating are similar to those of newts, the female picking up a spermatophore deposited by the male. A little later the female lays 10 to 70 eggs, each of which is ⅓ inch (1 cm) across, over a period of 3 weeks. Both male and female guard the eggs while they incubate; the parent coils its body around the eggs and gently waves its tail, presumably to increase oxygenation and to prevent silt from settling on them. After 4 months the larvae hatch. Almost 1 inch (2.5 cm) long, they lack limbs and gills and spend most of their time lying on their sides. Later they become more active and feed on minute algae and crustaceans.

Although almost black when they hatch, olms gradually lose this coloration and become a translucent white.

OLM

CLASS **Amphibia**

ORDER **Caudata**

FAMILY **Proteidae**

GENUS AND SPECIES ***Proteus anguinus***

LENGTH
Usually 8–12 in. (20–30 cm)

DISTINCTIVE FEATURES
Elongated, cylindrical body; broad head; compressed tail; limbs poorly developed with 3 toes on forefeet, 2 toes on hind feet; eyes small, usually concealed by skin folds; almost black (newly hatched), becoming translucent white (at rest) or pale pink (during activity); large, red, feathery gills on each side of neck

DIET
Small water-dwelling and mud-dwelling crustaceans and mollusks

BREEDING
Age at first breeding: 10 years; number of eggs: 10 to 70, guarded by both male and female; hatching period: about 4 months. Alternative (but rare) breeding strategy: fully developed young born live after retention of eggs inside female's body; number of young: 1 to 2.

LIFE SPAN
Up to 25 years

HABITAT
Subterranean lakes and streams in limestone caves; rarely above ground but sometimes found in streams after cave floods

DISTRIBUTION
Northeastern Italy and coastal regions of former Yugoslavia from Istria south to Montenegro

STATUS
Vulnerable

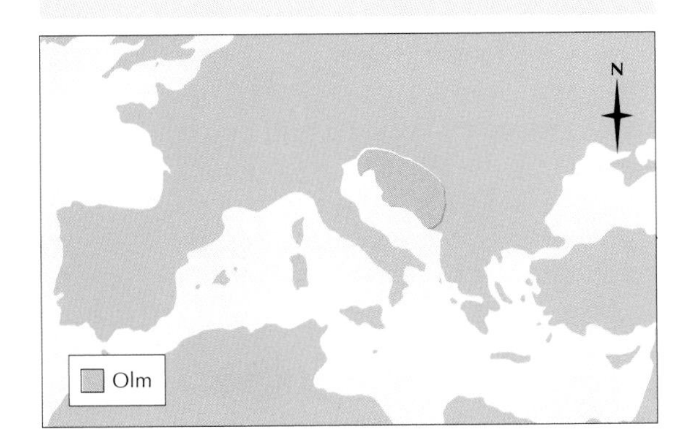

When the olm was first discovered, scientists' understanding of the relationships of animals was less extensive than it is today. The olm was wrongly thought to be a fish-lizard or even a fish-newt.

When olms have been kept in water warmer than the 41–50° F (5–10° C) of that of their native caves, they occasionally give birth to one or two live young, the eggs being retained in the female's body until they hatch.

Newly hatched olms are almost black, but they gradually lose the pigment as they develop, unless they are kept in the light, in which case the color is retained. Olms' eyes are also well developed to begin with, but they gradually degenerate and sink into the skin, becoming invisible, although the eye muscles and nerves remain. Olms mature at about 10 years and live for at least 25 years.

Refusing to change

Like the mudpuppy and the axolotl, *Ambystoma mexicanum*, the olm is an amphibian that retains its gills throughout its life and reproduces in the larval stage—a process known as neoteny. When it was discovered that an axolotl could, under certain circumstances, change into a salamander, experiments were carried out on olms to see if they could be induced to change. They all failed, so it seems likely that the olm is not the neotenic larva of another amphibian, as an axolotl is a neotenic salamander, but is a wholly aquatic amphibian that has never lived on land.

In this case, the olm is a primitive member of the order Caudata, or tailed amphibians. Originally the caudates kept their gills all their lives; but then modern types of caudates arose which spent part of their lives on land, such as the newts and salamanders. The olm is, therefore, an evolutionary relic, or missing link.

OPAH

Also known as the spotted moonfish, the opah is a large and highly distinctive fish. It hunts fast-moving prey in the midwaters of the world's oceans.

THERE ARE TWO SPECIES of opah, *Lampris guttatus* and the rare southern opah, *L. immaculatus*. Together, these fish represent the only members of the family Lampridae. The opah was formerly known as *Lampris luna*, meaning "radiant moon." This was presumably a reference to the fish's rounded body shape, although the opah is actually oval in outline.

The opah's body is strongly flattened from side to side, up to 6½ feet (2 m) or more long, and the fish weighs up to 595 pounds (270 kg). It has a tiny toothless mouth and its eyes are gold in color. The single dorsal fin runs from the middle of the back nearly to the tail; it is low for most of its length but rises into a high lobe in front. The anal fin is long and low. Both dorsal and anal fins can be folded down into grooves. The pectoral fins are high up on the flanks, and above them the lateral line rises in a high curve. The pelvic fins, like the pectorals, are scythe-shaped. The tail, which carries the tail fin, is a short stalk, sharply set off from the rear end of the body.

The opah's colors range from a deep blue on the back to a lighter blue, or silver, on the belly. The fish's scales are salmon pink in color and are iridescent, varying their shade according to the angle at which the light hits them. The fins are a vivid scarlet and there is scarlet around the mouth. Light-colored or silvery spots or flecks are scattered over the body.

The opah is an oceanic wandering species and is often found with tuna and billfish. It lives in the warmer parts of the Atlantic and Pacific Oceans, rarely entering the Mediterranean, but in summer it moves into temperate waters, going as far north in the Atlantic as Newfoundland, Iceland and Norway. In the Pacific it ranges as far north as Canada.

Feeds on fast movers

The opah is considered to be a midwater species, living at depths of 330–1,310 feet (100–400 m). However, it must spend a significant amount of time at or near the surface because that is where most opahs have been caught. The opah's rounded shape and considerable weight would seem to indicate that it is a slow-moving fish, and that, by implication, its prey is also slow moving. However, in his studies of the opah carried out during the mid-20th century, the naturalist P. A. Orkin

OPAH

CLASS **Osteichthyes**

ORDER **Lampridiformes**

FAMILY **Lampridae**

GENUS AND SPECIES **Common opah, *Lampris guttatus* (detailed below); southern opah, *L. immaculatus***

ALTERNATIVE NAME
Spotted moonfish

WEIGHT
Up to 595 lb. (270 kg)

LENGTH
Up to 6½ ft. (2 m)

DISTINCTIVE FEATURES
Strongly compressed, oval-shaped body; tiny, protrusible, toothless mouth; single dorsal fin runs from middle of back to tail; scythe-shaped pelvic and pectoral fins; deep blue back; lighter blue or silver belly with pink wash; white or silvery spots all over body; bright scarlet fins

DIET
Mainly other midwater fish and invertebrates, particularly squid and cuttlefish

BREEDING
Very poorly known. Breeding season: probably spring.

LIFE SPAN
Not known

HABITAT
Seas and oceans, generally at depth of 330–1,310 ft. (100–400 m)

DISTRIBUTION
Pacific: Canada south to southern Chile; Kamchatka to New Zealand; Atlantic: Nova Scotia and Norway south to Uruguay and Angola; Mediterranean; Indian Ocean

STATUS
Not threatened

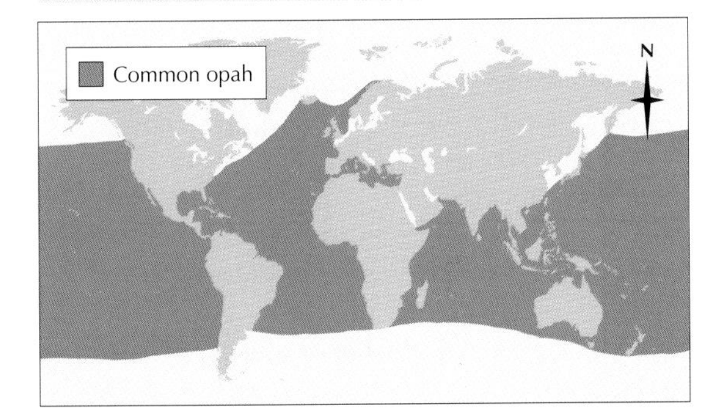

found that it feeds largely on small oceanic cuttlefish and squid, belonging to two genera, *Ommastrephes* and *Onychoteuthis*. These are among the fastest and most agile of almost any species of marine animal. Orkin's findings confirmed previous reports based on examination of the stomach contents of opahs, which also showed that crustaceans, especially isopods (marine wood lice) and small fish are taken. Orkin found as many as 50 squid beaks in the stomach of a single opah.

The stomachs of the fish Orkin examined contained the bones and otoliths (ear bones) of two fish that live at considerable depths. Another unexpected discovery in the stomachs of the opah was a littoral clam, *Donax variabilis*. Orkin was unable to explain how a fish living in a midwater region had been able to feed on this bottom-living animal.

When the opah opens its mouth, the bones of the upper jaw are projected forward, in the manner of that of the John dory, *Zeus faber* (covered elsewhere in this encyclopedia). This considerably enlarges the cavity of the mouth and allows the opah to suck in its prey. Another characteristic of the opah, and one that may be related to the fish's speed, is that the pectoral fins project horizontally from the sides of the fish. When it swims the opah is believed to move these fins up and down, in a similar manner to a bird beating its wings, which is an unusual use of the pectoral fins.

The opah does not move in schools and so is not caught in large numbers. However, it is considered a tasty food fish, the flesh being oily and having a delicate flavor similar to that of tuna, and the opah is caught year-round. The fish was regarded as a good luck symbol by longline fishers in the past, who would give it away as a goodwill gesture rather than sell it.

Uncertain breeding habits

Scientists have not been able to discover details of the opah's breeding habits. Despite the fact that the fish is found in many of the world's oceans, its breeding grounds remain unknown. Eggs taken from the roe of one captured female were measured at 0.8 millimeters in diameter. The eggs taken from another female, one that appeared to have finished laying, implying that the eggs left in the ovary were probably fully ripe, were 2.5 millimeters in diameter. The smallest opah ever caught was taken off the Pacific coast of North America and measured 23 inches (57.5 cm) in length. The next smallest on record was captured off Bordeaux, France, and measured 25 inches (62.5 cm). However, to date, scientists have no information about the opah between the egg stage and the young adult.

OPOSSUM

A pair of Virginia opossums. Opossums have strong toes and a flexible tail, enabling them to scale trees with ease to hunt for prey or escape from predators.

THE OPOSSUMS ARE THE only marsupials that live outside Australasia, except for the little-known caenolestids or rat opossums of the order Paucituberculata, which live in South America. The best-known opossum is the Virginia opossum, *Didelphis virginiana*, which is common in many parts of the United States and ranges into South America. The Virginia opossum is often known simply as the "possum," but this name is also, confusingly, given to Australian marsupials of the family Phalangeridae, such as the brush-tailed possums, genus *Trichosurus*, covered elsewhere in this encyclopedia. This is why it is now usual to refer to the American species as "opossum" and the Australian animals as "possum."

Virginia opossums are the size of a small dog, with a head and body length of 12–20 inches (30–50 cm), but their appearance is ratlike, with short legs and a pointed muzzle. The tail is almost as long as the body and is naked for most of its length. The ears are also hairless and the rough fur varies from black to brown or white. The hind feet bear some resemblance to human hands. The first toe is clawless and opposable in exactly the same way as the human thumb.

The other opossums that live in Central America and South America are similar to the Virginia opossum in appearance. Some have a bushy tail, and the water opossum, or yapok, *Chironectes minimus*, has webbed hind feet and a waterproof pouch. Most species are not known at all well, except for the common mouse or murine opossum, *Marmosa murina*, which sometimes damages banana and mango crops and is occasionally found in consignments of bananas.

Resilient marsupial

The spread of the Virginia opossum from the southeastern United States north to Ontario in Canada is remarkable for an animal that originated in tropical and subtropical climates. It is even more remarkable as it belongs to the marsupials, a group that in most parts of the world has become extinct in the face of competition from the placental mammals. The opossum's spread may be due to a decrease in the number of its predators, many of which have been killed off by humans. It is surprising that opossums have survived in the northern parts of their expanded range, as they are vulnerable to the cold and are sometimes found with parts of their ears or tail lost through frostbite. Although they do not hibernate, opossums become inactive in very cold spells, subsisting on fat stored during the fall.

Opossums generally live in wooded country, where they forage on the ground, climbing trees only to escape predators and to find food. They can, however, climb well, gripping with their opposable toes, while their tail is nearly prehensile. A young opossum can hang from a branch with its tail, but adults can use their tail only as brakes or as a fifth hand for extra support.

Each opossum has a home range of about 6–7 acres (2.4–2.8 ha), although it is sometimes twice as large as this. They feed mainly at night and

OPOSSUMS

CLASS Mammalia

ORDER Didelphimorpha

FAMILY Didelphidae

GENUS 15 genera

SPECIES About 70, including Virginia opossum, *Didelphis virginiana*; water opossum, *Chironectes minimus*; and common mouse opossum, *Marmosa murina*

ALTERNATIVE NAMES
D. virginiana: common opossum; *C. minimus*: yapok; *M. murina*: murine opossum

LENGTH
Head and body: 12¾–19¾ in. (32.5–50 cm); tail: 4–21 in. (10–53 cm);

DISTINCTIVE FEATURES
Ratlike body; large, forward-facing eyes; pointed muzzle; long, almost naked tail; most species gray brown in color

DIET
Mainly leaves, shoots, buds, seeds and invertebrates; some species also small vertebrates, carrion and garbage

BREEDING
Genus *Didelphis*. Age at first breeding: 1 year; breeding season: spring and summer; number of young: up to 20; gestation period: about 12 days; breeding interval: usually 1 or 2 litters per year.

LIFE SPAN
Up to about 3 years

HABITAT
Generally forest, wooded country and scrub; *C. minimus*: lakes and streams

DISTRIBUTION
Southern Canada south to Argentina

STATUS
Some species threatened; others common

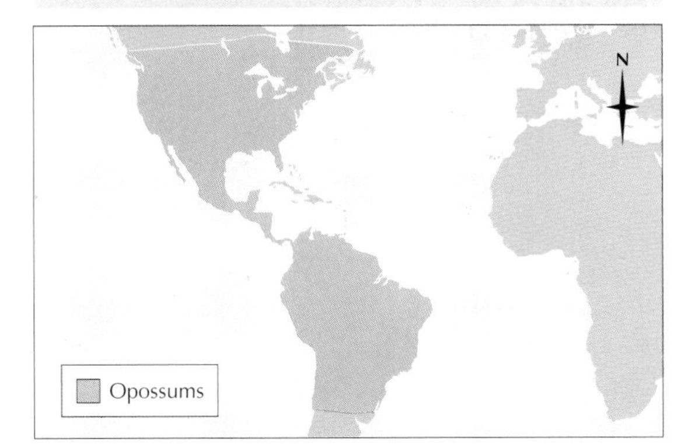

The Virginia opossum's fur consists of a thick undercoat concealed beneath longer, white-tipped protective hairs.

spend the day in a nest in hollow tree trunks, in abandoned burrows or under piles of dead brushwood. The nest is made of dead leaves that are carried in a most unusual way: they are picked up in the mouth and passed between the front legs to be held against the belly by the tail, which is folded under the body.

Varied diet

Opossums have sometimes been described as scavengers, mainly because they are often found feeding on garbage around human habitations. This simply provides an example of their adaptability, for they will eat a very wide variety of foods. Small ground animals such as earthworms, grasshoppers, beetles, ants, snails and toads are taken in large numbers in the summer and fall, together with voles, mice, snakes and small birds. Opossums also raid poultry runs from time to time and take plants such as pokeberries and persimmon, usually at night. They are particularly likely to take plant food during the late fall and winter, when their usual animal prey is becoming scarce, although they sometimes feed on carrion.

Opossums are preyed upon by many animals, including dogs, bobcats, coyotes, foxes, hawks and owls. They are also commercially trapped for their coats, which are of poor quality but are used to make simulated beaver or nutria (coypu) fur.

Carrying the babies

After a very short gestation of 12–13 days, the tiny young, numbering 8 to 18 in the case of the Virginia opossum, emerge from the mother's body in quick succession and crawl into her pouch to grasp her nipples. As she usually has 13 nipples, some of the litter may not be able to gain access to the mother's milk and soon die. The young remain in the pouch for 10 weeks, and after this they sleep huddled together in the nest. When the mother goes out foraging, she carries the young clinging to her back. They are now the size of brown rats, *Rattus norvegicus*. If the mother has a large litter, she may find it difficult to walk. The young are weaned shortly afterward and become independent at about 14 weeks. Females breed before they are 1 year old, producing one litter in the northern part of their range but sometimes two in the south.

Playing possum

The phrase "playing possum" comes from the opossum's habit of feigning death when frightened. The habit is not confined to opossums and has also been noted in foxes, African ground squirrels and various snakes. When an opossum is confronted by a predator and cannot escape quickly, it turns at bay, hissing and growling and trying to attack. If the predator succeeds in grabbing and shaking it, the opossum suddenly collapses, rolling over with its eyes shut and its tongue lolling out as if it is dead. The attacker frequently loses interest in the opossum at this, presumably because many of the opossum's natural predators do not eat carrion. After its attacker has left, the opossum gradually recovers and is able to make its escape.

This ruse must be effective at persuading predators to leave opossums alone, otherwise the latter would only be playing into the predators' hands. However, there is still some scientific debate as to how the opossum manages to simulate death so effectively. It has been suggested that harmless paralyzing chemicals are automatically released into the opossum's brain in reaction to high stress levels. Scientists believe that these cause the opossum's muscles to contract, producing an effect similar to rigor mortis. According to this theory, when the danger has passed the chemicals diffuse and the opossum recovers.

More recently, experiments have been made on opossums using an electroencephalogram. This machine records the patterns of minute electric currents in the brain, showing differences between waking and sleeping states. Recordings made of opossums feigning death showed that far from being in a state of catalepsy, the marsupials are wide awake and alert.

The common mouse, or murine, opossum, found only in South America, is one of the smallest opossums.

OPOSSUM SHRIMP

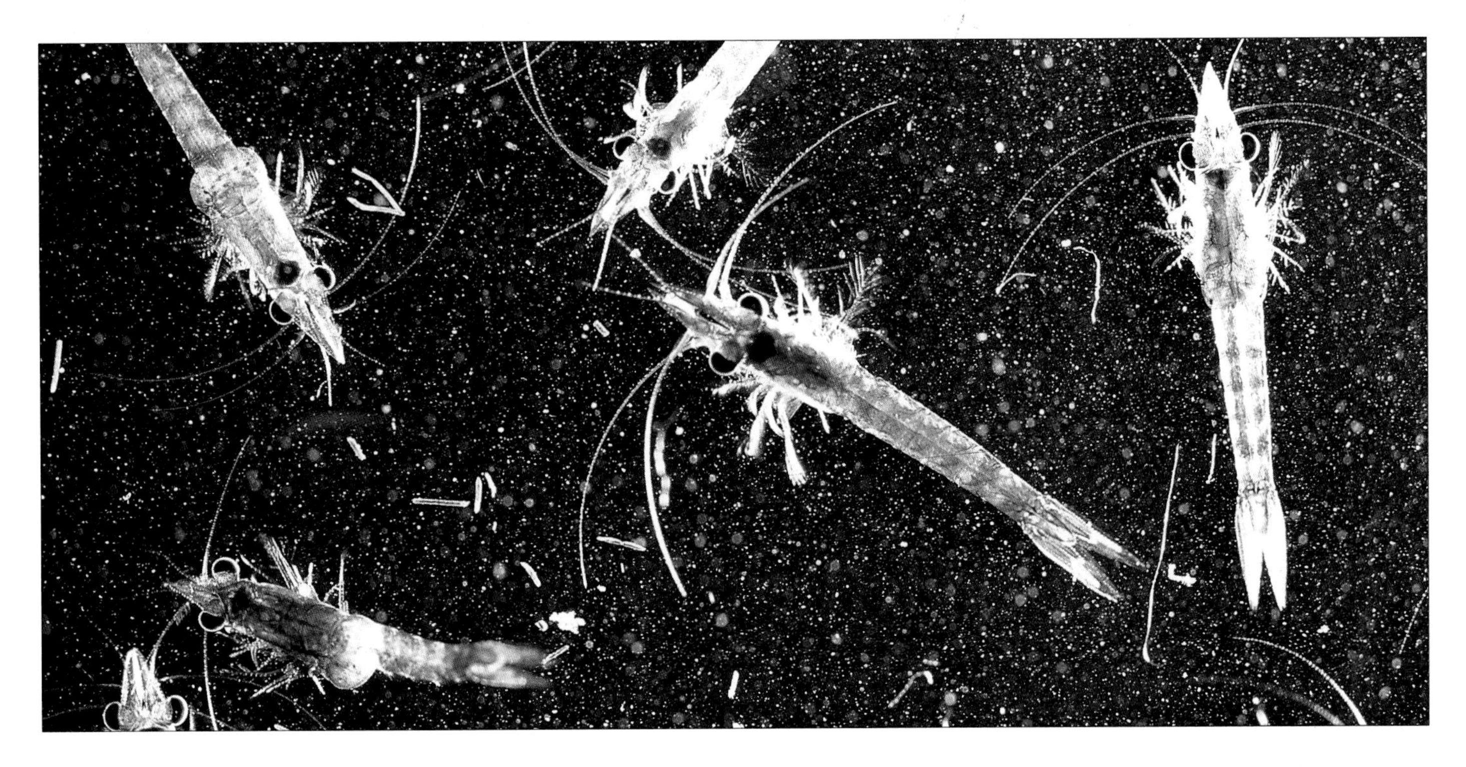

Opossum shrimps are very important as food for marine fish and are also sold for human consumption in East Asia. **Neomysis integer** *(above) is one of the most abundant European species of opossum shrimps.*

THE OPOSSUM SHRIMPS ARE small shrimplike crustaceans, the females of which, rather like opossums, have a brood pouch, or marsupium, to hold their eggs and young. The name opossum shrimp is often applied to all the species in the order Mysidacea—some 650 of them—or it may be reserved for those of the genus *Mysis*. Most of the Mysidacea are marine, although about 25 species live in fresh water and twice that number are found in brackish water.

The typically shrimplike body is divided into a thorax and an abdomen. The thorax consists of eight segments covered, except at the rear, by a carapace fused with the head and the first three segments. The abdomen, made up of six segments, ends in a tail fan, which bears, in most species, a pair of balancing organs characteristic of the group. These organs, called statocysts, are little chambers. In each of them, on a group of sensory hairs, sits the statolith, a mass that consists partly of calcium fluoride. Opossum shrimps have two pairs of antennae on the head. The stalked and movable compound eyes are absent in a few cave-dwelling species. Most opossum shrimps are ⅕–1 inch (5–25 mm) long, although some are as short as 3 millimeters or as long as 7½ inches (19 cm). As a general rule, the length increases from shallow water to deep water and from the Tropics to the poles.

The color of opossum shrimps varies from species to species, although probably most are transparent, or nearly so. Opossum shrimps may be deep red in the deep seas or bright green in shallow waters, perhaps to match a bright green seaweed. The color is inclined to vary with the background, with which the opossum shrimps tend to harmonize. Color changes are caused by the movement of pigments, called chromatophores, inside groups of cells. A false red color is often the result of food inside the gut being visible through the transparent carapace.

Opossum shrimps are denser than water and must keep swimming if they are not to sink. Some species swim with their thoracic limbs and some with the swimmerets on the abdomen. While some species are entirely pelagic, most swim near the bottom by day, often tending to rise toward the surface at night. When disturbed, opossum shrimps can spring backward by flexing the abdomen and tail fan under the thorax and suddenly straightening it again, sometimes so powerfully that they jump right out of the water. *Gastrosaccus spinifer* half walks and half swims over the bottom and rapidly buries itself in sand if disturbed.

Filter feeders

Most opossum shrimps feed by filtering fine particles of dead plants and animals as well as microscopic organisms from the water. Larger foods, living or dead, may also be eaten by some species. The filter feeders draw water currents in toward the mouth by the vibration of appendages on the head called maxillae, each of which

OPOSSUM SHRIMP

PHYLUM **Arthropoda**

CLASS **Crustacea**

SUBCLASS **Malacostraca**

ORDER **Mysidacea**

FAMILY **Mysidae**

GENUS AND SPECIES ***Neomysis integer***

LENGTH
Up to ⅔ in. (1.7 cm)

DISTINCTIVE FEATURES
Small, shrimplike crustacean. Cylindrical body, divided into thorax and abdomen; 2 pairs of antennae on head; abdomen ends in tail fan; generally transparent.

DIET
Particles of dead plants and animals; also microscopic living organisms

BREEDING
Breeding season: just after female has molted; hatching period: a few weeks; eggs and young carried in female's brood pouch

LIFE SPAN
Up to 2 years

HABITAT
Shallow pools and waters to depth of 66 ft. (20 m), occasionally much deeper; on open coasts and in estuaries

DISTRIBUTION
European coasts

STATUS
Superabundant

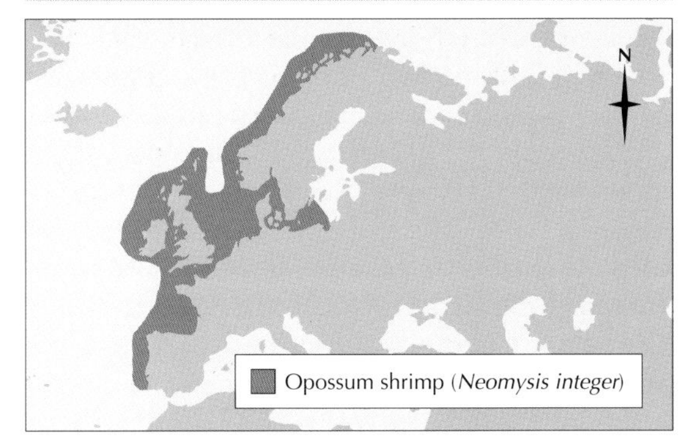

Some opossum shrimps breathe through gills located on the thorax or the abdomen. Others have no gills but breathe through their carapace, which is rich in blood vessels.

ends in a filtering comb of bristles. In many species the maxillae are helped by the beating of the limbs on the thorax.

Young safe in a brood pouch

The breeding habits are known for only a few of the opossum shrimps. The most characteristic feature, from which they get their name, is the protection of the young in a brood pouch on the underside of the female's thorax. During mating, the male approaches the female and places sperm into the brood pouch through a long genital tube. In some species, he simply sheds his sperm and lets water currents carry the sperm into the brood pouch. Then, 20–30 minutes later, the female lays her eggs in the brood pouch, where they are fertilized. Usually two to several dozen eggs are laid. The young stay in the pouch for a few days to a few months after hatching, depending on the temperature and species. They are released during the evening or at night, and the female molts and mates again the same night.

Ice Age relicts

The few opossum shrimps that live in fresh water are restricted to localities near the coast, where they are subject to sea spray. The most widely distributed freshwater species is *Mysis relicta*, which is ½–1 inch (1.5–2.5 cm) long and found in certain freshwater lakes in northern Europe and North America and in the brackish waters of the Baltic Sea. It is a relict from the changes that took place in the last Ice Ages. The theory is that the species arose in a saltwater lake formed when part of the Arctic Ocean was impounded by glaciation. As the water became fresher, the shrimps adapted to the change and gave rise to *M. relicta*. Subsequently, they spread from lake to lake along the retreating ice margin.

ORANGUTAN

THE ORANGUTAN IS ONE of our most interesting relatives. It occupies an intermediate place within the Hominoidea, the superfamily that comprises the apes and humans, because it is less closely related to human beings than the gorilla or chimpanzee, but more humanlike than the 11 species of gibbons.

A large male orangutan stands up to 5⅗ feet (1.7 m) high when upright and may weigh as much 200 pounds (90 kg). Females stand only 4 feet (1.2 m) at the most and on average weigh half as much as the males. The orangutan's arms are 1½ times as long as the legs. Both the hands and feet are long and narrow and suited for grasping, and the thumb and great toe are very short. The orangutan's skin is coarse and dark gray, and the hair, which is rufous orange in color, is shaggy but sparse, so the skin can be seen through it in many places. The male develops large cheek flanges or pads and grows a beard or mustache, the rest of the face being virtually hairless. There is a great deal of variation in facial appearance between different animals. Orangutans are as individual and instantly recognizable as human beings. Both sexes have a laryngeal pouch, which in the male can be quite large, giving him a flabby appearance on the neck and chest. The forehead is high and rounded, and the jaws are prominent. Youngsters have a blue tinge to the face.

Orangutans are found on Borneo, an island of the Malay Archipelago, and on Sumatra in Indonesia. There are slight differences between the two subspecies, and these are more marked in the male. The Bornean subspecies, *Pongo pygmaeus pygmaeus*, is maroon tinted, and the male has enormous cheek flanges and great dewlaps formed by the laryngeal sac. The Sumatran subspecies, *P. p. abelii*, is larger and lighter colored, and males can look quite startlingly human, with only small flanges and sac, a long narrow face and a long, gingery mustache.

Old man of the woods

The orangutan is strictly a tropical forest animal. It generally lives in low-lying, even swampy forests, but is also found at 6,000 feet (1,830 m) on mountains in Borneo. Here, at any rate, most individuals are entirely arboreal (tree-living). They swing from branch to branch by their arms, although they may use their feet as well or walk upright along a branch, steadying themselves with their hands around the branch above. Sometimes old males become too heavy to live in the trees, and so spend more time on the ground.

A young orangutan in Kalimantan, Borneo. The Bornean subspecies is smaller than the Sumatran orangutan and has a darker coat with a maroon tint.

When they are on the ground, orangutans move quadrupedally, with the feet bent inward and clenched and the hands either clenched or flat on the ground. This contrasts with the gorilla and chimpanzee, which live mainly on the ground and "knuckle-walk," with their feet flat on the ground and their hands supported on their knuckles. In captivity, orangutans easily discover how to walk erect, but because the leg muscles are insufficiently developed, the knee is kept locked and the leg straight.

Forest loners

At night the orangutan makes a nest between 30 and 70 feet (9–21 m) above the ground. It is often built with a kind of sheltering roof to protect the animal from the rain, a structure that is not found in nests made by chimps or gorillas. The nest is otherwise much more loosely made than those of chimps or gorillas. It takes only 5 minutes to make and the orangutan usually moves on and makes a new nest at its next night's stop-

Every day in late afternoon orangutans take about 5 minutes to build a simple nest high up in the trees of their forest habitat.

ping place. Sometimes the same one is used again and the previous night's nest may be used for a daytime nap.

Unlike gorillas and chimpanzees, orangutans seem to have no large social groupings. A female with her infant often travels with other such females for a while, forming something like a smaller version of the chimpanzee's nursery group. A male may join this group, but adult males live alone most of the time. Adolescents of both sexes tend to travel around in groups of twos or threes.

Large vocabulary of sounds

It is thought that male orangutans, like gibbon families, may be territorial, spacing themselves by means of calls. The laryngeal sac is filled with air, swelling up enormously, the air then being released to produce what has been described as a loud, two-tone booming sound. Orangutans communicate within a group by making a smacking sound with their lips every few seconds. They are also able to produce a loud roar. This begins on a high note and the tone gets deeper and deeper as the laryngeal sac fills with air. Roaring can be heard at night and before dawn, and orangutans are also said to make the same noise before mating. The Dayaks (the Indonesian peoples of interior Borneo) have reported that fights between male orangutans and scars are quite common.

ORANGUTAN

CLASS **Mammalia**

ORDER **Primates**

FAMILY **Pongidae**

GENUS AND SPECIES **Bornean orangutan, *Pongo pongo pygmaeus*; Sumatran orangutan, *P. p. abelii***

WEIGHT
66–200 lb. (30–90 kg); adult male about twice as heavy as adult female

LENGTH
Head and body: 4–5⅗ ft. (1.2–1.7 m)

DISTINCTIVE FEATURES
Body resembles that of large chimp; arms 1½ times longer than legs; long, narrow hands and feet with short thumb and great toe; sparse, shaggy coat of rufous orange hair. Adult: large cheek flanges (pads), particularly in male; male also has beard or mustache.

DIET
Fruits and other plant matter; also insects, small vertebrates and bird eggs

BREEDING
Age at first breeding: 7 years (female), rarely before 14 years (male); breeding season: all year; number of young: usually 1; gestation period: about 245 days; breeding interval: 3–6 years

LIFE SPAN
Up to 60 years in captivity

HABITAT
Tropical rain forest

DISTRIBUTION
Borneo and Sumatra

STATUS
Endangered; estimated population in early 1990s: about 15,000 (Sumatra), up to 20,000 (Borneo)

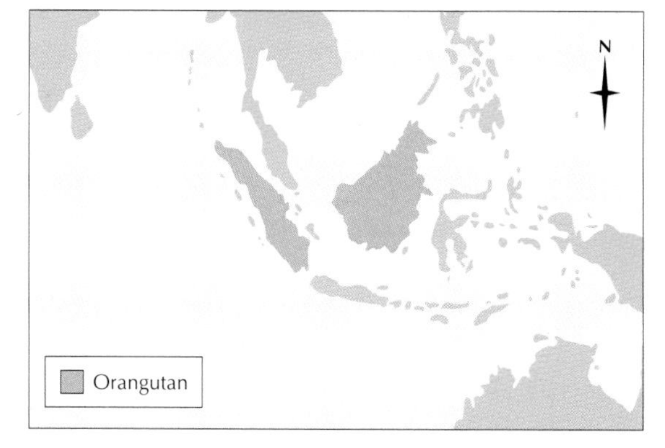

Small, almost naked young

There is no special birth season, food being available year-round in the rain forests of both Sumatra and Borneo. Gestation lasts 245 days and the single young, occasionally twins, weighs only 2–3 pounds (1–1.4 kg) at birth. It is sparsely covered with hairs on the back and head. At first the young orangutan clings to its mother's fur, usually slung on her hip, but when it is a little older it wanders about on its own, sometimes walking along the branch behind its mother, clinging to her rump hairs. At about 4 or 5 years of age the young leave their mothers and form adolescent bands.

Few natural predators

Humans are the main threat to the orangutan. Orangutans love the juicy, evil-smelling durian fruit, and so do human beings, so this is often a source of contention. An orangutan will react to a human intruder by making a great deal of the noise and breaking off branches. It will keep up a continuous shower of branches and foliage, which may be annoying enough to drive away the intruders. There are very few natural predators of the organgutan. Tigers were once a threat, but there are no longer any tigers on Borneo, and only a few remain on Sumatra.

Shrinking distribution

The orangutan's distribution has been steadily declining in size. Its ancestors' remains have been found in 14-million-year-old deposits in the Siwalik Hills, Punjab, India. In the Pleistocene, 500,000 years ago, the orangutan was found as far north as China and as far south as Java. Today it occurs in small areas of primary rain forest in the states of Subah, Sarawak and Kalimantan in Borneo and only in the north of Sumatra. It seems that deforestation and ever-increasing human populations have seriously affected this animal's distribution and there are now fears that it may become extinct altogether in the wild. A slow breeding rate has been another aspect of the orangutan's decline. There is a 3–6 year gap between births, and a female does not usually breed again until the previous young has dispersed. It is possible that a typical female bears only three or four babies in her life.

Pet and zoo trade

After habitat loss, the biggest threat to the orangutan's survival is the zoo and pet trade. Every zoo wants a young ape to display to its visitors, and orangutans are the easiest to obtain. Many unscrupulous private zoos, especially in the United States, have paid high prices for baby orangutans, and there has been a lucrative trade in them in Southeast Asia. Baby orangutans also make popular pets among the rich of some countries, only to be later abandoned or shot when they become too large and strong to keep at home. The babies are obtained by shooting their mothers. Most later die in transit, often being transported in appalling conditions. For every one orangutan that reaches a zoo alive, 10 have probably died.

It is now illegal to catch, sell or kill an orangutan and there are fines to deter poachers. However, the trade continues and each year hundreds of orangutans are taken from their rain forest homes. Orangutans are now classified as endangered. In addition to strict protection in the species' remaining habitats, conservation methods include rehabilitation centers for orphaned young and captive breeding in zoos. In some cases animals have been successfully released back into the wild.

A large mature male orangutan. Despite strict laws prohibiting the capture and sale of wild orangutans, an illegal trade in the animals still flourishes in many parts of Southeast Asia.

ORB SPIDER

A lesser garden spider (*Meta segmentata*) in its web. Mating in this species is a perilous business for the male, which waits until the female is busy feeding before surreptitiously mating with her.

Orb spider is a collective name for those spiders belonging to the family Araneidae that spin an orb-shaped web, sometimes called a cartwheel or geometric web. The vast majority of these spiders, with their bodies large by comparison with their legs, are entirely dependent on their webs to catch their prey. One of the most numerous and best studied orb spiders is the European garden spider, also known as the diadem or cross spider.

The female is ½ inch (1.2 cm) long, excluding the legs, and the male is about half this, although the sizes reached by males are variable. These spiders vary greatly in appearance, with their colors ranging from a drab pale fawn to a rich brown. They often have patterning on the abdomen, which may be marked with spots, blotches and lines in white, yellow or shades of brown. The most characteristic of the markings is, in its simplest form, a group of five white spots forming a cross. This pattern may be extended into a series of spots or oval markings.

The wonders of the web

All spiders live by eating other animals, predominantly insects. Most of them construct some form of web made of silk threads, which are given out from a group of spinnerets on the underside of the abdomen. In orb spiders the web starts with an outer scaffolding. The first thread laid down is known as the bridge thread. It is horizontal, and side threads are joined to it to complete a rectanglular, polygonal or triangular frame. A series of radial threads are then laid down from the frame to a hub near the center. At the hub a closely woven platform is made from which a spiral of temporary scaffolding is run outward. All this is made of nonsticky silk. Using the nonsticky threads to walk on, the spider, working from the outside inward, lays down a close spiral of sticky threads, distinguished by the beads along each thread. As the spider is making the sticky snare, it cuts away the temporary scaffolding as it goes. The hub is often placed just above the geometric center of the web. This is because the spider can move faster downward due to gravity, and by placing the hub off center the spider can still reach all parts of the web quickly. Despite this, the spider often has a better chance of catching prey before it escapes from the bottom half of the web, and in many species this is reflected by the greater number of sticky threads in this area.

It is a popular idea that the silk is liquid when first given out of the spinnerets and that it hardens on contact with the air. In fact, it coagulates as it is being squirted out and becomes strong and extensible as the spider pulls on it.

For their size, webs are exceptionally strong; larger webs can trap small birds and bats. Orb spiders of the genus *Nephila*, for example, spin

ORB SPIDERS

PHYLUM **Arthropoda**

CLASS **Arachnida**

SUBORDER **Labidognatha**

FAMILY **Araneidae**

GENUS ***Meta, Argiope, Araneus, Nephila, Micrathena, Singa*; many others**

SPECIES **About 4,000**

ALTERNATIVE NAME
Orb-web spider

LENGTH
Up to 2 in. (5 cm)

DISTINCTIVE FEATURES
Large, colorful abdomen, often with angular projections or spines; 8 relatively short legs; 8 eyes, with middle 4 often forming a square; often waits at hub of elaborate orb web

DIET
Small invertebrates, especially flying insects

BREEDING
Number of eggs: 4 to 1,200, depending on species and size of spider; silk egg sacs attached to web, leaf litter or plants

LIFE SPAN
Most species: 1 year or less; several other species: up to 9 years

HABITAT
Very varied; includes gardens, woodland, forest, grassland and meadows

DISTRIBUTION
Virtually worldwide

STATUS
Most species common; many abundant

webs that may be as long as 8 feet (2.5 m) across, and which are made of thick, tough silk. In parts of Southeast Asia local people have used these webs as fishing nets, bending a pliable stick into a loop with a handle and passing this through a web so that it comes away attached to the loop.

Many species of orb spiders have stabilimenta on their webs; these are bands of thickened silk that zigzag diagonally across the webs with the spiders in the center. The silk bands make the webs clearly visible, which may seem to be counter-productive for the spiders. However, the stabilimenta glow brightly in ultraviolet light, which is visible to insects. Many types of flowers have ultraviolet guides on their petals that show pollinating insects where to go. The orb spiders' stabilimenta mimic these ultraviolet guides, fooling flies and other insects into thinking that the deadly webs are merely flowers. Birds, which are important predators of orb spiders, can also see in ultraviolet light, and the stabilimenta may therefore also serve to make the spiders look larger and more dangerous.

When an insect flies into a web and begins to struggle, the orb spider is alerted by the vibrations and runs out to secure its victim, usually swathing it in silk. At the same time the spider injects poison into its victim from the chelicerae (fangs). The poison not only paralyzes the prey but also acts as a digestive juice, liquefying the contents of the body. Later the spider inserts its fangs into the victim's body and uses them as tubes for sucking out the nutrient-rich liquid.

Not all orb spiders make webs. Bolas spiders catch their prey by swinging a single thread with a sticky globule at the end above the abdomen.

An* Araneus *orb spider from Western Australia. When orb spiders spin a web they start with an outer scaffolding of nonsticky silk threads, then work inward, creating a hub at the center. Finally they add a tight spiral of deadly sticky threads.

Silk leaving the spinnerets of an* Argiope *orb spider. This remarkably tough material hardens yet remains elastic as it leaves the body because of changes in the pH (acidity) of the silk.

Some bolas spiders release chemicals into the air, mimicking the chemicals emitted by female moths, thus attracting the male moths to their doom.

When spiderlings leave home

The eggs of the European garden spider are laid in autumn in 1½ inch (4 cm) long silken cocoons of a dingy golden yellow color. Each cocoon contains 600 to 800 eggs. The cocoons are fastened in a secluded, dry place, and the eggs hatch the following June. As soon as they hatch, the spiderlings spin an irregular mass of very thin strands of silk in which they cluster in a ball. When disturbed, they rush around in all directions, coming together again once things have settled down. They stay together in this way for a few days and then start to disperse.

When a spiderling arrives at a suitable site, it spins a small orb web about 2 inches (5 cm) across, similar to the familiar cartwheel shape but much more irregular. At this time there is little difference between male and female, but after a while the females begin to grow more quickly. They do not mature until the following summer, and then growth in size is very rapid, the females outstripping the males and making webs up to 2 feet (0.6 m) across. This explains why the orb webs seem suddenly to appear in late summer. The spiders have been there all the time, but they and their webs have suddenly grown larger. The majority do not survive the first frosts and by the end of November most of the adults have died, although a few hide away in sheltered spots, including in houses.

Climate the worst enemy

In temperate latitudes, the weather, especially drought and excessive rain, is the orb spiders' worst enemy. Also, many small birds include orb spiders in their diet, and the spiders are also preyed upon by insects such as ants. Parasitic wasps paralyze some spiders before laying their eggs on the spiders' bodies and burying the corpses. After hatching, the young wasps feed on the still-living spiders. Some orb spiders prey on their own kind, and occasionally a female eats the male after copulation.

The power of silk

As well as being elastic, orb spider silk is phenomenally strong. More energy is required to snap spider silk than a similarly sized length of high-tensile steel. Biotechnologists are looking at ways of using spider silk in a wide range of products, including parachute cords, surgical sutures and bullet-proof vests.

ORIBI

THE ORIBI IS ONE OF the small, straight-horned dwarf antelopes and is closely related to the gazelles. It is among the largest of this group, 20–28 inches (50–71 cm) at the shoulder, with a silky coat, tufts of hair on the knees and a bushy black tail. It is sandy red to brown above and white below. There are white partial rings around the eyes and a gland in the corner of each eye. Rather like the reed-bucks (genus *Redunca*), the oribi also has a scent gland below each ear, located beneath a dark, bare patch of skin.

The oribi is found in sub-Saharan Africa from the Sudan south to South Africa, and from central Ethiopia in the East to Senegal in the West. The species was abundant until the spread of European settlement but is now extinct in many areas.

Living in small groups

Oribis live in open grassland, on rolling downs and in bush country. They are diurnal (day-active) and prefer to feed in the early morning, often coming out before sunrise and then lying up in the grass after 6–7 a.m. Around noon oribis can often be seen standing or lying in tall grass or lying up in hollows. They usually go around in pairs, but groups of up to six animals have been seen.

Series of mating gestures

The rutting (mating) season is in the early part of the year. Gestation is about 210 days, and the young are born from September to January according to the part of Africa in which the parents live. Oribis set up territories, each pair or family party living in an area of ½–1 square mile (1.3–2.6 sq km). They mark grass stalks around the boundaries of their territories with waxy secretions from their eye glands to keep other oribis away. When the female defecates or urinates, the male stamps in the excrement, so presumably transferring scent from the glands between his front hooves. All this imparts an individual atmosphere to the territory, and if one pair are removed, another pair will move in within a day or two.

During courtship the male raises his foreleg and strokes the female's hind leg. This action is called the *lauf-schlag* gesture and is typical of gazelles and their relatives. The male follows the female as she walks slowly along with tail held erect. Occasionally the male rises on his hind legs as if to mount, but the female continues walking.

After some 15 minutes the male stretches his head forward, pushes it between the female's hind legs and lifts her rear end off the ground. He trundles her forward on her front legs for a few paces. He then withdraws his head, examines and sniffs her genitalia and pushes his head under again. It seems that the act of pushing under is the direct stimulus to mating, since once the male has repeated this procedure two or three times, the female stands still, allowing the male to mount her and mate with her.

When the young oribi is born, the female eats the placenta so it does not attract predators. As do all small antelopes, the oribi fawn grows quickly, doubling the birth weight after 3 weeks. The fawn hides in the long grass, and the mother moves away from it as a diversionary tactic at any sign of danger. The young animal reaches its full size at about 8 months.

An oribi rests amid boulders and tall grass during the heat of the day. The oribi is found at altitudes of up to 10,000 feet (3,000 m) and can go for a long time without drinking.

The oribi is equipped with a range of scent glands, the secretions from each of which serve a different purpose. For example, scent secreted from the ear glands, and fanned by the ears, identifies an individual oribi.

Running and jumping

Oribis are small animals and are preyed on by lions and even small cats, as well as by pythons, eagles and baboons. They sleep lightly and wake easily to the shrill whistling alarm call, galloping off with a stiff gait, rumps bobbing up and down. As the oribi runs, its tail is held stiffly up, wagging. There are frequent stotting episodes, behavior seen in many gazelles and dwarf antelopes, in which the animal jumps into the air with all four legs straight and stiff. Stotting gives the animal a better view and warns others visually of danger and allows them to follow. It also shows a would-be predator that the stotting animal is fit and would be hard to catch. After dashing off 100 yards (90 m) or so, the oribis stop and turn to watch the source of the disturbance, fleeing again if danger still threatens.

One possible reason for the great decrease in oribi numbers lies in their faithfulness to their territories. When disturbed, they tend to circle back to the place from which they started, so presenting an easy target. They are also inquisitive and approach unfamiliar objects, even an armed human, advancing cautiously, stopping, then advancing again. When a fire sweeps through their territory, the pair flee but return behind the wall of flames, presumably to subsist on pockets of vegetation left by the fire or on quick-growing grasses and herbs that shoot up through the ashes.

ORIBI

CLASS **Mammalia**

ORDER **Artiodactyla**

FAMILY **Bovidae**

GENUS AND SPECIES ***Ourebia ourebia***

WEIGHT
31–46 lb. (14–21 kg)

LENGTH
Head and body: 36–55 in. (0.9–1.4 m); shoulder height: 20–28 in. (50–71 cm); tail: 2⅓–6 in. (6–15 cm)

DISTINCTIVE FEATURES
Small size; sandy red to brown upperparts; white belly, chin and inner thighs; tuft of hair on each knee; black, bushy tail; short, straight horns (male only)

DIET
Mainly grasses and leaves of low bushes

BREEDING
Age at first breeding: 10–15 months; breeding season: all year, but dependent on climate; number of young: 1; gestation period: about 210 days; breeding interval: usually 7–9 months

LIFE SPAN
Up to 12 years

HABITAT
Savanna and bush, close to wooded copses

DISTRIBUTION
Sub-Saharan Africa

STATUS
Locally common; scarce in some areas

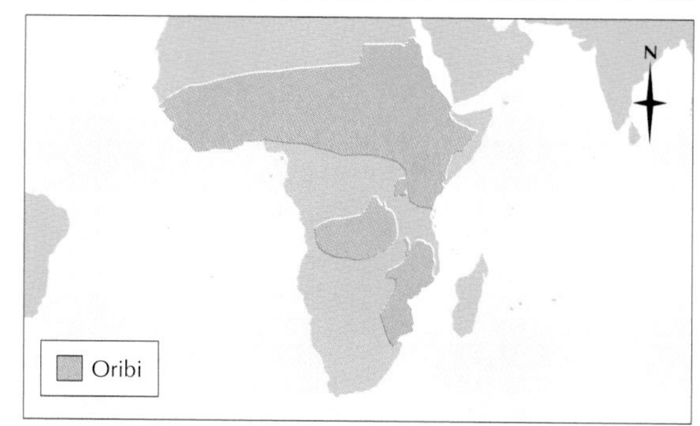

OROPENDOLA

The oropendolas belong to the Icteridae, a family that includes such diverse birds as the orioles, American blackbirds, cowbirds, caciques, grackles, bobolinks, military starlings and meadowlarks. The family ranges throughout the Americas, the bobolink (*Dolichonyx oryzivorous*) nesting in Canada and the austral blackbird (*Curaeus curaeus*) in Tierra del Fuego in southernmost South America, but most icterids, including the oropendolas, are tropical.

Probably the best-known oropendola is the Montezuma oropendola, *Psarocolius montezuma*, which is found from Mexico south to Panama. It is the largest of the genus, about the size of an American crow. The males are 18–20 inches (46–51 cm) long, while the females are slightly smaller. The plumage is very striking, with a black head, neck and chest, a chestnut body and a brownish black tail with bright yellow outer feathers. The long sharp bill is black at the base and red at the tip. In the male the base of the bill is edged with orange and there is an orange wattle on each side of the chin. The bill extends over the forehead, giving the oropendola a streamlined appearance. There is a patch of bare skin, white tinged with blue, on each cheek.

Oropendolas live in the Central American forests, where they can sometimes be seen flying in large straggling flocks searching for food in the tops of the trees or in clearings, and along the banks of rivers and pools. Their flight is like that of crows, with a steady, measured beat. The males sing throughout the year, bowing from a perch with the bill below the level of the feet. The song is made up of a large number of very different phrases, as is usual with the songs of icterids. From a distance it is pleasant to hear, but from nearby rasping sounds like rusty machinery can be heard among the continual liquid gurgling and chattering notes. Males also make a characteristic noise as they fly. This sound is caused by vibrations of the stiff parts of their flight feathers.

Diet of fruits and nectar

Oropendolas are mainly fruit eaters, but they also eat insects and drink nectar. There are very occasional reports of them catching and eating smaller birds, such as small species of grosbeaks. They forage among the forest canopy for soft fruits and berries but often come into plantations, where they eat ripe bananas. Oropendolas

A pair of chestnut-headed oropendolas, **Psarocolius wagleri,** *at the birds' nest site. Oropendolas weave hanging sacklike nests.*

Oropendolas have long pointed bills that extend onto the crown, giving the birds an unusual, streamlined appearance. Pictured is a russet-backed oropendola, Psarocolius angustifrons*, in the Venezuelan rain forest.*

also hang upside down, poking their long bills deep into the banana flowers in order to drink the nectar inside.

Weave hanging nests

Some of the North American icterids are referred to as hangnests after their sacklike nests that are suspended from twigs. The nests of oropendolas and caciques hanging like pendants from the outer branches of trees are a remarkable sight. The nest of the Montezuma oropendola consists of a long purse of neatly woven fibers, sometimes up to 6 feet (1.8 m) long, that hangs from slender branches and twigs or from a fork. The entrance is at the top of the purse, and in its position in the outer reaches of the forest canopy the nest is well protected from predators. Oropendolas are, however, parasitized by cowbirds.

The female oropendolas do most of the work when it comes to building nests, while the males perch nearby singing and keeping watch. Observation of these birds at close quarters is often difficult because the males give alarm calls when possible danger approaches, sending the females scuttling into the foliage. To build her nest, a female oropendola nicks the underside of the midrib of a leaf with her bill and then tears off a strip of fibers up to 2 feet (60 cm) long. The first step in nest-building is to make a foundation around the chosen twig and then to construct a ring of fibers attached to this. The ring forms the entrance of the nest, and the oropendola weaves the main body of the nest from this as if knitting a sock. She invariably works from inside the

MONTEZUMA OROPENDOLA

CLASS **Aves**

ORDER **Passeriformes**

FAMILY **Icteridae**

GENUS AND SPECIES ***Psarocolius montezuma***

WEIGHT
Male: ¾–1¼ lb. (350–530 g); female: ½–⅔ lb. (200–250 g)

LENGTH
Head to tail: male 18–20 in. (46–51 cm); female 15–16 in. (38–41 cm)

DISTINCTIVE FEATURES
Large size; long, pointed bill, top of which extends to crown; mainly chestnut plumage with black head, neck and chest; naked patch of pale blue-white skin on head; orange wattle on chin (male only); bright yellow outer tail feathers

DIET
Mainly fruits, berries and large insects; occasionally flower nectar

BREEDING
Age at first breeding: 1 year or more; breeding season; January–May; number of eggs: 2; incubation period: about 15 days; fledging period: about 30 days; breeding interval: 1 year, but high-ranking males breed with several females each year

LIFE SPAN
Not known

HABITAT
Mainly forest edge and coffee plantations

DISTRIBUTION
Southeastern Mexico south through Central America to southern Panama

STATUS
Common

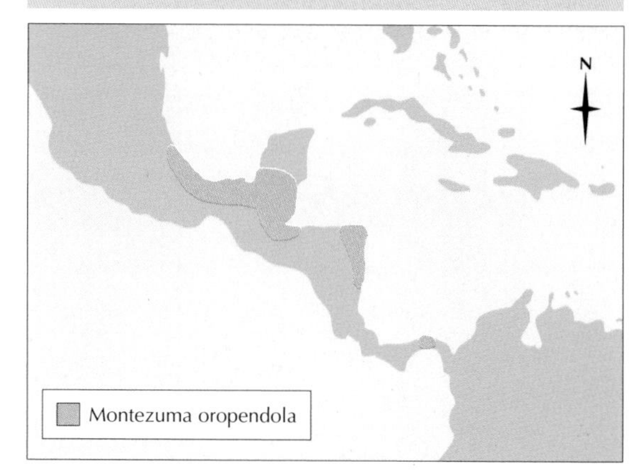

growing nest, always entering and leaving by the entrance rather than by the open, unfinished end of the "sock." The nest ends as a round inverted dome, and finally it is lined with leaves.

Oropendolas live in colonies where the nests may be so tightly packed that they are woven together. Sometimes a twig or branch snaps under the combined weight of nests, broods and sitting parents. This means that oropendolas are likely to have pieces of their nests stolen by neighbors while they are away searching for more material, but there are advantages to communal living. It often happens that a nest is not securely anchored and attachment to another nest may prevent it from falling. Also, many eyes are more likely to spot predators than those of only one pair of birds, and the chances of an individual bird being preyed upon or parasitized are also much reduced.

Females outnumber males by about six to one in each colony. An alpha male is at the top of the hierarchy within a colony, and does most of the mating. He also has first pick of the local food resources, as he is heavier than lower ranking birds. These lower ranking males do mate, but they do it away from the colony. Fighting between males is rare, but when it does occur it can be severe. The males take no part in rearing the young, but they are present for the rest of the season, watching out for predators. Each female lays two eggs, which are white with speckles of black, and these are incubated for about 2 weeks. The chicks spend about a month in the nest being fed on fruit by the female.

Diversity of birds

The icterids are diverse not only in general appearance but also in their feeding and breeding habits and are a good example of adaptive radiation. It is not immediately obvious that the colonial oropendolas, with their long purse-like nests, belong to the same family as the parasitic cowbirds. Bobolinks and meadowlarks, meanwhile, build simple nests of grass stems near the ground. Their social life is conventional, although two or more females may sometimes be found with one male and the male does little to help in raising the brood.

In general, there is a link between the diet of an icterid and its breeding habits. The insect-eaters, such as the blackbirds and the caciques, are monogamous and nest solitarily, while the fruit-eaters, such as the oropendolas, are polygamous and colonial. There are, however, many variations within this framework, such as the cowbirds, which do not make nests at all.

The hanging nests of a breeding colony of chestnut-headed oropendolas, in the Panamanian forest. These intricate nests are built solely by the females, using strips of fibers torn from large leaves.

ORYX

Gemsbok (above) twist and turn when trying to evade predators. By contrast, Arabian oryx keep to a straight line. The latter also run less rapidly than gemsbok.

ORYX ARE AMONG THE most striking of antelopes, white or fawn in color, with long, back-pointing, ridged horns. Their closest relatives are the sable and roan antelopes and the addax. However, they differ from the first two species in that the horns sweep back in line with the face instead of rising vertically above the eyes, and from the addax in having smaller hooves and horns that do not spiral.

There are three species of oryx, all living in the desert areas of Africa and southwestern Asia. The dramatically colored gemsbok, *Oryx gazella*, is 4 feet (1.2 m) high and is fawn colored, with a black nose patch and eye stripe uniting to form a "bridle" around the muzzle. It has white shanks, a white belly, a black throat fringe and a black stripe down the back and along the flanks, which extends onto the fore- and hind legs.

The second species is the rare Arabian oryx, *O. leucoryx*. Small and white, it stands 3½ feet (1 m) tall and has dark brown or black face marks, a line down the throat and dark limbs.

The third and most distinctive species of oryx is the scimitar oryx, *O. dammah*, found in the Sahara Desert. It has distinctly back-curved horns and is white with reddish head markings and a red neck. The red color may or may not extend back over the body.

Three subspecies of gemsbok are also recognized by zoologists. These are the beisa oryx (*O. g. beisa*), the tufted oryx (*O. g. callotis*) and the true gemsbok (*O. g. gazella*).

Dangerous horns

These wary and extraordinarily keen-sighted animals wander in herds over large distances. When they are alarmed, the whole herd takes off at a gallop. The Arabian oryx moves rather slowly and clumsily compared to the gemsbok. When brought to bay, however, the oryx turns on its pursuer, jabbing and butting with its horns. There are reports of trucks, used in capturing oryx for conservation purposes, having been speared by oryx horns. During the rutting (mating) season the males fight fiercely but are protected from serious injury by the thickened skin on their shoulders.

Oryx herds are often small, ranging from half a dozen individuals to perhaps a dozen. They are composed mainly of females and young, most males living a solitary existence. In the rainy season, the small herds unite into larger groups of as many as two dozen oryx, sometimes up to 60. These herds contain many pregnant females. The herds feed mainly at night or in the early morning, on grasses, herbs, fruits and roots.

ORYX

CLASS **Mammalia**

ORDER **Artiodactyla**

FAMILY **Bovidae**

GENUS AND SPECIES **Scimitar oryx, *Oryx dammah*; gemsbok, *O. gazella*; Arabian oryx, *O. leucoryx***

ALTERNATIVE NAMES
***O. gazella beisa*: beisa oryx;**
***O. g. callotis*: tufted oryx**

WEIGHT
220–463 lb. (100–210 kg)

LENGTH
Head and body: 5–8 ft. (1.5–2.4 m); shoulder height: 3–4½ ft. (0.9–1.4 m)

DISTINCTIVE FEATURES
Slim, straight horns up to 5 ft. (1.5 m) long (both sexes); basic coloration variable, includes browns, grays and cream; striking black or dark brown markings; tufted tail

DIET
Grasses, herbs, roots, tubers and fruits

BREEDING
Age at first breeding: 1–3½ years; breeding season: year-round; number of young: 1; gestation period: about 240 days; breeding interval: 1 year in favorable conditions

LIFE SPAN
Up to about 20 years

HABITAT
Arid plains, thick bush and rocky hillsides

DISTRIBUTION
***O. gazella*: southwestern and East Africa; *O. leucoryx*: reintroduced to a few sites in Oman, Jordan and Saudi Arabia; *O. dammah*: restricted to Chad; almost extinct in wild**

STATUS
***O. gazella*: scarce; *O. leucoryx*: endangered; *O. dammah*: critically endangered**

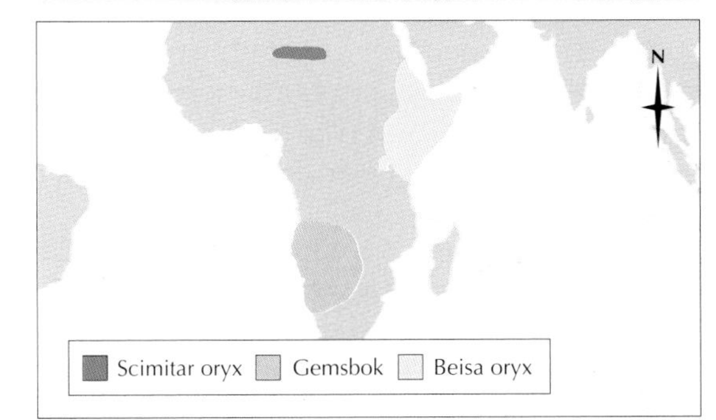

An Arabian oryx in Saudi Arabia. Oryx are primarily grazers, but browse if grass is unavailable. They can survive without water by digging up roots or tubers for moisture.

Ceremonious courtship

The inherent aggressiveness of the oryx finds unexpected expression in their courtship behavior. The male and female spar, head-to-head with horns crossed, pushing with their heads, and moving from side to side, forcing each other around in circles. Then the male rubs the female's hindquarters with his cheek and lays his head on her back. Raising his foreleg horizontally, he strokes the outside of her hind leg or between her hind legs. This movement, known as the *laufschlag*, is a common feature of the courtship of gazelles, ibex and many other bovids. At the same time, the male's head and body are stretched upward, and the female lowers her head. Both the sparring and the *laufschlag* alternate several times. This behavior differs from the sparring between two males in that the partners are much closer together, too close to do each other any harm, and the heads are lowered instead of raised. The female may not want to mate, and, if so, she will attempt to "defeat" the male. The couple may whirl around over 20 times. Then the female runs away and the male eagerly pursues her. If the female is willing to mate, the pair circle around four or five times. Finally, the male chases the female, mounts her, clasping her loins with his forelegs, and holds his head and neck high while the

A herd of beisa oryx in Kenya. Herds are generally composed of females and young.

female's head is lowered. As in all artiodactyls (even-toed hoofed mammals), copulation itself takes only a second or two.

The gestation period is about 240 days. Calves are often born in the rainy season, when there is plenty for them and their mothers to eat. In Arabia this period lasts from May–September, but in the Kalahari, where the true gemsbok is found, it lasts from September to January.

Keeping cool

The oryx, which lives in desert country, must go without water for long stretches and travels during the dry season into more fertile areas. The scimitar oryx, for example, moves out of the Sahara, as far south as 15° N, although conditions there are still extremely hot and arid. Some moisture can be obtained by eating plants with a high water content or by feeding on roots or tubers for their moisture. However, the full explanation of how the oryx manages to survive in such conditions was not known until 1969, when the zoologist C. R. Taylor published the results of some of his latest experiments.

When the air temperature is higher than an animal's body temperature, the animal heats up. To avoid overheating it loses this extra heat by sweating or panting. The oryx lives in an environment that is both hot and dry. It needs to lose heat, but if it sweats it will lose valuable water that cannot be replaced. Taylor found that the oryx can tolerate a considerable rise in body temperature. Its normal body temperature is 96° F (35° C). In an experiment to test the animal's capacity to withstand heat increases, the surrounding air temperature was raised to 104° F (40° C). In response, the oryx's body temperature rose 12 Fahrenheit degrees to 108° F (42° C) before evaporation of sweat began to increase, preventing any further temperature rise. With the body temperature greater than that of the air, heat was also lost by conduction and radiation. However, provided the oryx had access to water, its temperature usually increased by only half as much before sweating accelerated.

When deprived of water, the oryx does not waste the supply in its body, but lets its body temperature rise. In this way, not only does the oryx not absorb heat, it actually lose it. This inevitably places a strain on the body, so if water is available, the oryx uses it for evaporation. An oryx's brain, the part that is most sensitive to overheating, remains cool when the body temperature goes up. This is because the carotid artery, the artery in the neck that carries blood to the brain, branches into a network of vessels. These are close enough to veins, bringing cooler blood from the nose, that they lose heat. The heat is passed from artery to vein, cooling the blood before it passes to the delicate brain. Moreover, the oryx feeds at night, when the relative humidity of the desert air is increased by the lower temperature. Food plants that are dry and crumbly by day, containing as little as 1 percent water, absorb moisture at night, and within 10 hours contain 42 percent water.

Operation Oryx

The rare Arabian oryx once ranged throughout Arabia into Syria and Iraq, but by 1962 it was nearly extinct in the wild, although there were a small number in Oman and the Federation of South Arabia. In the 1940s and 1950s intensive hunting killed many, and there were fears for the species' survival. Some Arabian oryx were captured and sent to the Phoenix Maytag Zoo, Arizona, where there is a desert environment similar to that of Arabia. The British Fauna Preservation Society mounted Operation Oryx in April–May 1962 and succeeded in capturing two males and one female. These were flown to Phoenix, where they were joined by a female from the London Zoo and one donated by the Sultan of Kuwait. They were bred successfully and there were then 18 in all, two of which were sent to the Los Angeles Zoo to form a second breeding group. In 1977 there were over 100 in captivity, 94 of which were in the United States.

The Arabian oryx became extinct in the wild in 1972. However, several were reintroduced into Oman in 1982, into Jordan in 1983 and into Saudi Arabia in 1990. There are now about 500 animals in the wild, and about 2,300 in captivity.

OSPREY

An osprey soars over Florida, where there is a year-round resident population. Ospreys tend to nest high in trees or on rocks, but in Florida they also nest on the ground beside highways.

THIS REMARKABLE BIRD OF prey, which lives almost exclusively on fish, is a highly efficient hunter, with up to 90 percent of its dives being successful. The osprey, also known as the fish hawk in North America, is the size of a small eagle, and in flight its long, narrow wings can span more than 5 feet (1.5 m). Its large toes have long, curved claws, and the undersides of the toes are covered with spiny scales that help to hold the fish. The head and short crest are white with a black band running through the eye to join the chocolate brown plumage of the back. The underparts are white except for a dark band across the breast. The plumage of the sexes is similar, but the females are larger.

The osprey is almost cosmopolitan. It breeds in North America, northern Asia and much of China and northern and eastern Europe, as well as in scattered places such as southern Spain and the coasts of the Red Sea. In the Southern Hemisphere it breeds regularly only in Australia and adjacent islands to the north, but it often migrates to South America and South Africa.

In the higher latitudes of the Northern Hemisphere the osprey is migratory. Birds nesting in northern Europe overwinter in sub-Saharan Africa, returning the next spring. Migratory ospreys usually follow rivers or coasts but occasionally fly cross-country. Many die en route, especially young birds. Ospreys can cover great distances quickly. One migrating bird was satellite-tracked from Ireland to Spain in September 2000. In a 36-hour period it flew 626 miles (1,008 km) across the Atlantic and Bay of Biscay. In their winter range ospreys may have territories that they defend against other ospreys.

Spectacular diving

Ospreys spend most of their time circling over water or perched on rocks or trees where there is a good view over the water. They are usually found on the coast, but they also haunt lakes and rivers. In Australia they live on the coral islets of the Great Barrier Reef.

When it hunts, this spectacular fisher circles over the water on its long wings at a height of 50–100 feet (15–30 m) until it spots a fish. It hovers for a moment, then plunges, hitting the water with a great splash. It then surfaces triumphantly to bear its catch back to a perch. The osprey appears to strike the water headfirst, but just before the final impact, it throws its feet forward and enters the water talons first, grabbing the fish with both feet, one in front of the other. Sometimes its dive is so violent that it disappears beneath the water, but at other times

The osprey's feet are well adapted for catching fish. Not only do they have horny spines, or spicules, on the underside of the toes, but the outer toe on each foot can be turned until it faces backward, improving the grip still further.

it descends gently from the air to pick up its prey from the surface. The angle of the bird's dive is usually about 45°, but it can be almost horizontal.

The fish the osprey catches are usually those, such as pike, that bask near the surface, but bream, carp, perch, roach and trout are also frequently caught, depending on the locality. They are taken at a depth of no more than 40 inches (1 m). One osprey fishing at sea was found to eat mainly needlefish. On rare occasions ospreys have been found to eat mice, beetles, wounded birds and even chickens, but these items are probably taken only when the ospreys are very hungry.

There is a remarkable record of an osprey that met its death from being too good a fisher: a carp netted in a lake in Germany had the skeleton of an osprey firmly attached to its back. Presumably the carp, which weighed almost 10 pounds (4.5 kg), pulled the osprey underwater and drowned it. The talons were so deeply embedded that the corpse could not be freed, so it decomposed while trailing behind the carp.

Where they are common, ospreys nest in colonies, with the nests sometimes as little as 180 feet (55 m) apart. The nest is constructed mainly

OSPREY

CLASS **Aves**

ORDER **Accipitriformes**

FAMILY **Pandionidae**

GENUS AND SPECIES ***Pandion haliaetus***

ALTERNATIVE NAME
Fish hawk (North America only)

WEIGHT
2½–4½ lb. (1.1–2 kg)

LENGTH
Head to tail: 21⅔–22¾ in. (55–58 cm); wingspan: 57–67 in. (1.45–1.7 m)

DISTINCTIVE FEATURES
Large size; long, relatively narrow wings; powerful, strongly hooked bill; long flight feathers form "fingers" at tips of wings; long, strong, feathered legs; bluish feet; white below; chocolate brown above; white head with black line through eye

DIET
Almost exclusively fish

BREEDING
Age at first breeding: 2–3 years; breeding season: April–May (North America and northern Europe), virtually all year (rest of range); number of eggs: 2 or 3; incubation period: 34–40 days; fledging period: 49–56 days; breeding interval: 1 year

LIFE SPAN
Up to 30 years

HABITAT
Sea coasts, estuaries, rivers, inland lakes and marshes with open water

DISTRIBUTION
Breeds in much of North America, northern Eurasia, eastern China, North Africa, the Middle East, Southeast Asia and Australasia

STATUS
Locally common

of seaweed, heather, moss, sticks and dead branches, sometimes stripped off trees while the bird is in flight. Ospreys usually nest in trees or among rocks but may build on the ground. Some have been persuaded to build their nests on artificial platforms such as cartwheels set on stakes. Occasionally ospreys become a nuisance by building on telephone poles and pylons.

In Europe and North America, the usual clutch of two or three white eggs with brown markings is laid in late April or May. The female does most of the incubation, which lasts 34–40 days, and during the first 30 days of the chicks' lives she stays in the nest, brooding or shading them. During this period all her food is brought to her by the male. At first the chicks are fed with small lumps of semidigested fish, but later they are given raw strips and when 6 weeks old they are left to tear up fish for themselves. They make their first flights at 7–8 weeks, and they either learn to catch their own fish or perish. Many ospreys die in their first year.

The eggs and chicks fall prey to nest-robbers, especially when the parents leave the nest through being disturbed. Crows and raccoons are known to steal eggs, and eagles and gulls may do so. The parents often try to lure potential predators away from their nest with a distraction display. They utter loud calls and stagger about in the air with their feet dangling, making themselves conspicuous and taking attention away from the nest.

Ospreys in Britain

Ospreys were once fairly abundant in the British Isles but were almost exterminated by the increase in the numbers of shooting and fishing estates patrolled by gamekeepers. The birds were finally wiped out by human egg collectors. In 1954 the ospreys made a comeback in Scotland. Conservationists thwarted the efforts of nest-robbers, and numbers of ospreys breeding in Scotland continued to increase. In 2000, 120 pairs of ospreys nested in Scotland.

An adult osprey watches from its nest built at the top of a tree in Scotland. An osprey nest is about 6½ feet (2 m) across and is reused year after year.

OSTRICH

THE OSTRICH IS THE largest living bird and is also one of the most familiar because of its unusual appearance. A large male may stand over 9 feet (2.7 m) high, and almost half of this great height is neck. The male's plumage is black except for the white plumes on the wings and tail. The popularity of these plumes as adornments led to ostrich numbers being greatly reduced in many places and later to ostriches being raised on farms. The female's plumage is brown with pale edging to the feathers. The head, most of the neck and the legs are almost naked, but the eyelids have long, black lashes. There are two strong toes on each foot, the longer being armed with a large claw. Despite being flightless, ostriches can cover long distances by running quickly on long, strong legs.

A few million years ago, in the Pliocene era, there were several species of ostriches, but only one survives today. About 200 years ago five subspecies of this species ranged over much of Africa, Syria and Arabia, in desert and bush regions. They are now extinct or very rare over most of this large range. For example, the Arabian subspecies, *S. c. syriacus*, was last positively recorded in 1941. However, ostriches are still plentiful in East Africa, and also live wild in areas of South Africa and a few parts of southern Australia, where they were introduced.

Two female ostriches at the nest shared by a male's harem. The huge eggs are creamy white and almost round. Their tough shells are about 2 millimeters thick.

Stranger than fiction

Ostriches are extremely wary, their long necks enabling them to detect disturbances from quite a distance. As a result, it is very difficult to study ostriches in the wild, and until recently scientists' knowledge has been based mainly on observations of domesticated ostriches. Incomplete observations in the wild have led to many mistaken ideas about the habits of these birds, which have now become legendary. A husband and wife team of zoologists, E. G. F. Sauer and E. M. Sauer, studied ostriches in southern Africa by the ingenious method of disguising their blind as a termite mound. Ostriches and several other animals treated this blind with complete indifference, with the result that the Sauers were afforded a grandstand view of ostrich social life, and they found that in some respects this is almost as strange as the legends.

Ostriches often live in very dry areas, and they move about in search of food, often in fairly large herds. The herd is led by a male or female, which chooses grazing grounds and makes decisions about when to move. If the herd leaves familiar territory or comes to a water hole where no other animals are drinking, the dominant ostriches push the immature birds forward to spring any ambushes. During wet spells the herds break up into family groups, consisting of a pair with chicks and immatures.

Ostriches feed mainly on plants, including fruits, seeds and leaves. In deserts they obtain water from succulent plants. Ostriches also eat small animals, including lizards and small tortoises. Their reputation for eating almost anything, including lumps of metal and tins of paint, is widespread and perhaps exaggerated, but ostriches swallow considerable amounts of sand to aid digestion, and it is said that it is possible to trace the movements of an ostrich by examining the kinds of sand and gravel in its stomach.

Flexible ostrich society

Until recently there was doubt over whether ostriches were polygamous or monogamous. Those who believed they were monogamous pointed out that there was never more than one male or one female seen at a nest or leading a group of chicks. It is now known that ostriches may be monogamous in areas where food is scarce, but more usually

OSTRICH

CLASS **Aves**

ORDER **Struthioniformes**

FAMILY **Struthionidae**

GENUS AND SPECIES ***Struthio camelus***

WEIGHT
Male: 220–285 lb. (100–130 kg); female: 200–245 lb. (90–110 kg)

LENGTH
Height to top of head. Male: 83–108 in. (2.1–2.75 m); female 69–75 in. (1.75–1.9 m).

DISTINCTIVE FEATURES
Massive; extremely long neck; huge, strong legs; only 2 toes on each foot; neck and legs pink, gray or blue, according to subspecies. Male: mainly black plumage; white plumes on wings and tail. Female: dull brown.

DIET
Mainly plant matter such as leaves, stems and seeds; also some invertebrates, small lizards and small tortoises

BREEDING
Age at first breeding: 3–4 years; breeding season: all year, depending on locality; number of eggs: 4 to 8 per female and up to 25 per nest (several females usually share same nest); incubation period: 39–42 days; fledging period: 120–150 days; breeding interval: about 1 year

LIFE SPAN
Up to 40 years

HABITAT
Savanna and semidesert

DISTRIBUTION
Africa, from southern Sahara to Namibia; introduced to South Africa and Australia

STATUS
Common in a few areas but scarce over much of range

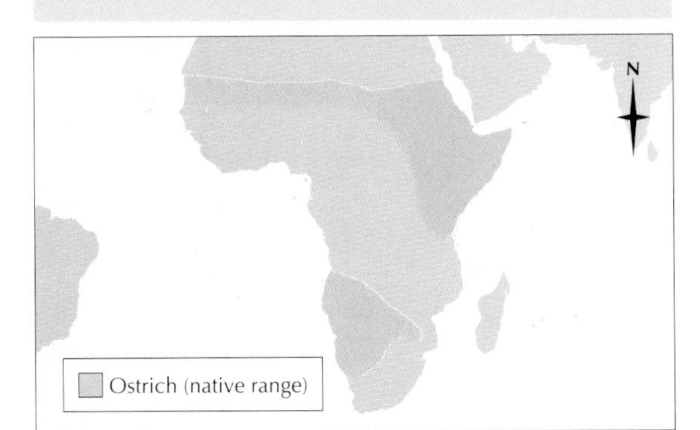

The ostrich's neck is covered in fine feathers, giving the appearance of being bare. Its color varies from pink to gray to blue, depending on subspecies.

they are polygamous. The Sauers found that the social organization of ostriches is very flexible and that a male accompanying a female with chicks need not be the chicks' father.

Breeding takes place at any time of the year, depending on when the rainy season occurs. At first the males develop a red coloration on their heads and feet and they display to each other, chasing around in groups with wings held out to show off the white plumes. Later they establish territories away from the communal feeding grounds, and here they are joined by the females. A male ostrich usually has three hens (females) in his harem, but it is not unknown for him to have as many as five.

The courtship ceremony is elaborate. The male separates one female from the harem and the pair feed together, synchronizing the movements of head and neck. The male then sits down and opens his wings to show the white plumes. At the same time he rocks from side to side and twists his neck in a corkscrew. The female walks around him and eventually drops into the mating position.

Each female lays four to eight eggs, which are about 6 inches (15 cm) long and weigh up to 2½ pounds (1.1 kg). The members of a harem all lay in one nest, which consists of a depression in the ground that may be 9 feet (2.7 m) across. It can take nearly 3 weeks for all the eggs to be laid, after which the dominant hen drives the others away and guards the nest with the male. One nest in Kenya was found to contain 78 eggs, but only 21 were incubated. The dominant hen pushes the other hens' eggs toward the edges of the nest. Incubation consists of keeping the eggs cool by shading them rather than keeping them warm. Toward the end of the 39–42-day incubation period the most advanced eggs are rolled into pits on the edge of the nest. This is probably a mechanism to synchronize the hatching of the eggs as much as possible. The chicks can run almost as soon as they hatch and after a month are able to attain a speed of 30 miles per hour (48 km/h). When they leave their parents, they form large bands, breeding when 3–4 years old.

Running to safety

Adult ostriches have little to fear from predators. They are wary and can run at 40 miles per hour (64 km/h) or more in short bursts and at 35 miles per hour (56 km/h) over long periods. The eggs and young ostriches, on the other hand, may fall prey to jackals and other predators. The adults lead their chicks away from enemies and perform distraction displays while the chicks scatter and crouch. Beating their wings and calling loudly, the ostriches run to and fro, presenting a broadside to the predators and occasionally dropping to the ground and setting up a cloud of dust with the wings. The male may continue the display while the female leads the chicks away.

Burying their heads

One of the popular notions about ostriches is that they bury their heads in the sand when danger threatens. The idea goes back to Roman times at least, and like so many legends, there is a basis of truth in it. The story is probably due to the difficulty in observing ostriches. When an ostrich is sitting on the nest, its reaction to disturbance is to lower its head until the neck is held horizontally a few inches above the ground. The ostrich is then very inconspicuous and the small head may well be hidden behind a small plant or hummock. Also, distant ostriches bending down to feed may look as though they are sticking their heads into the ground.

An ostrich sprints over the desert. The long, powerful legs end in feet that have only two toes. This is unique in birds; all other species have three or four toes.

OTTER SHREW

THE LARGEST OF THE THREE species of otter shrews, all of which are African, resembles a miniature otter. It is known as the giant otter shrew, *Potamogale velox,* and is found in large, fast-flowing rivers and streams in rain forest areas, from Nigeria east to western Kenya. In some places the giant otter shrew lives in forest pools in the rainy season and migrates to rivers in the dry season.

Half-otter, half-shrew

The head and body length of the giant otter shrew can be up to 14 inches (35 cm) and it has a stout tail that grows up to 12 inches (30 cm). The tail tapers to the tip and is flattened laterally. As it swims, the otter shrew moves the tail with a fishlike, side-to-side motion. The smooth, sleek fur is made up of a short, soft, dense underfur with a covering of long, coarse guard hairs. The legs are short, with five toes on each foot, and unlike those of otters are relatively weak. The flattened head has a long muzzle, small ears and tiny eyes. Each nostril has a flap of skin that acts as a valve when the animal is underwater. The end of the snout is swollen, almost duck-billed in shape, and has many long whiskers that act as organs of touch. Scientists believe that most of the giant otter shrew's food is found through touch, particularly when it is hunting underwater. Although the whiskers are mainly tactile sensory organs, they also act like a hydrofoil for the otter shrew underwater, helping to stabilize its head and body as it swims.

Dwarf species

The two dwarf African otter shrews grow up to 14 inches (35 cm) long, including the tail, and are brownish gray above and gray below. The toes are partly webbed, and the tail, keeled above and below, is round and covered with short hair. There are two species of dwarf otter shrews, the Mount Nimba otter shrew, *Micropotamogale lamottei,* and the Ruwenzori otter shrew, *M. ruwenzorii.* The two dwarf otter shrews look more like true water shrews than otters. Like water shrews, and in contrast to the giant otter shrew, they swim by paddling with their feet.

Secretive habits

All three species of otter shrews have a very restricted distribution. The giant otter shrew lives in western and central tropical Africa, from

Like otters, the giant otter shrew has a streamlined body and powerful tail, both adaptations for swimming. However, unlike otters, its feet are weak and provide very little propulsion in the water.

sea level to 6,000 feet (1,830 m). The Mount Nimba otter shrew lives in the coastal regions of West Africa, while the Ruwenzori otter shrew lives in the Democratic Republic of the Congo.

The giant otter shrew swims rapidly and easily with a sculling action of the tail and serpentine movements of the body, legs pressed against the sides. On leaving the water, it often scratches its belly vigorously with the second and third toes on the hind foot and then licks its feet. Although it moves awkwardly on land, it can run at a fair speed.

Active by night

There is still scientific debate about the habits of otter shrews. From captured females, zoologists have surmised that there are probably up to four young, and that there are two breeding seasons per year. The giant otter shrew is a solitary animal, pairing only to breed. It becomes active 2–2½ hours after nightfall, and throughout the night until first light it alternates feeding with resting in irregular periods. During the day it shelters in burrows in the banks of streams. The entrance to a burrow is above water level and the tunnel runs for a long way into the bank and then forks. One branch leads to a sleeping chamber in which lies a nest of dry leaves. The other branch leads down and opens farther along the bank, underwater.

A freshwater diet

As with many aquatic animals, otter shrews probe in mud or under stones for food. They bring their prey onto land in order to eat it. The diet of otter shrews consists mainly of freshwater crabs, and an individual may consume 20 to 25 crabs per day. When eating a crab, an otter shrew first turns it onto its back, then tears out the flesh from the softer underside. Otter shrews also take a variety of fish, amphibians, frogs, aquatic insects and mollusks. The smaller Ruwenzori otter shrew preys largely on worms, insect larvae, tadpoles and small crabs.

Endangered species

All three species of otter shrews are endangered to some extent. Hundreds of giant otter shrews are trapped by hunters in Africa every year, and many others drown accidentally in fishing nets. Otter shrews sometimes get into fish traps and kill all the fish, then accidentally drown, as they are unable to escape. Numbers of the Mount Nimba otter shrew have declined as a result of mining operations in Guinea and Liberia, which have released pollutants into rivers and streams. Gold-panning in streams has also brought about a steep decline in the population of the Ruwenzori otter shrew.

OTTER SHREWS

CLASS **Mammalia**

ORDER **Insectivora**

FAMILY **Tenricidae**

GENUS AND SPECIES **Giant otter shrew, *Potamogale velox*; Mount Nimba otter shrew, *Micropotamogale lamottei*; Ruwenzori otter shrew, *M. ruwenzorii***

WEIGHT
***P. velox*: about 12¾ oz. (360 g). *M. lamottei* and *M. ruwenzorii*: about 4¾ oz. (135 g).**

LENGTH
***P. velox*. Head and body: 11½–14 in. (29–35 cm); tail: 9⅔–12 in. (24–30 cm). *M. lamottei* and *M. ruwenzorii*. Head and body: 4¾–8 in. (12–20 cm); tail: 4–6 in. (10–15 cm).**

DISTINCTIVE FEATURES
***P. velox*. Streamlined body and stout, broad tail give otterlike appearance; flattened head; chocolate brown above; whitish below. *M. lamottei* and *M. ruwenzorii*: resemble water shrews; brownish gray overall.**

DIET
Freshwater crabs, small fish, frogs, tadpoles, aquatic insects and worms

BREEDING
Age at first breeding: not known; breeding season: all year (*P. velox*); number of young: probably 1 to 4; gestation period: not known; breeding interval: 2 litters per year

LIFE SPAN
Probably less than 5 years

HABITAT
Rivers, streams and pools, mainly in forests

DISTRIBUTION
Small ranges in western and central Africa

STATUS
All species endangered

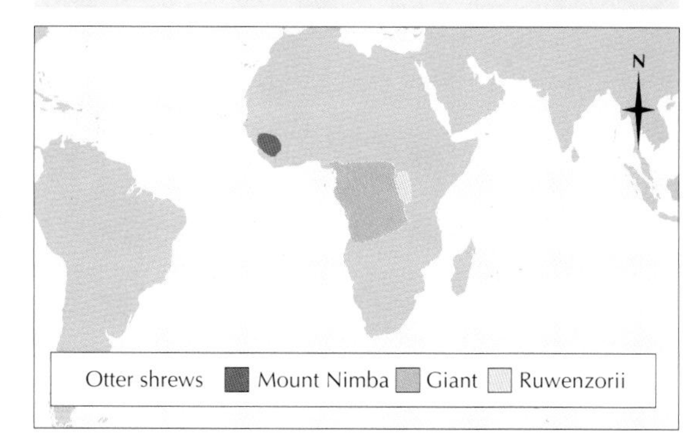

OVENBIRD

THE LARGE FAMILY OF South American ovenbirds is named after the shape of the clay nests made by the true ovenbirds, or horneros. The 240 species are divided into several groups of diverse forms, including the true ovenbirds of the genus *Furnarius*, the spinetails, thornbills, earth-creepers, shaketails, foliage-gleaners, leafscrapers and many others. Most members of the family have dull brown plumage with rufous or chestnut shades and vary from 5–11 inches (12.5–27.5 cm) long. The spinetails, genera *Synallaxis*, *Cranioleuca* and *Asthenes*, are sometimes known as castlebuilders because of the size of their nests. Spinetails have very long, sometimes forked tails and brighter plumage than most ovenbirds. The pale-breasted spinetail, *Synallaxis subpudica*, has bright yellow and black on its chin, while the Des Mur's wiretail, *Sylviorthorhynchus desmursii*, has a tail that is two or three times as long as the rest of the bird. In some species the webs of the tail feathers have become reduced so that the central quill projects. The 7–8-inch (17.5–20-cm) long rufous hornero, *F. rufus*, also known as the barber, is reddish brown above and whitish below, and is recognized as Argentina's national bird. The plumage is usually dull, but there is some variety in form between species. A few ovenbird species have crests, and the plain xenops, *Xenops minutus*, has an unusual upturned, wedge-shaped bill.

Ovenbirds range from southern Mexico to Patagonia, but the largest number of different forms are found in the Argentinian pampas and mountainous regions of Chile. In North America there is an ovenbird, *Seiurus aurocapillus*, that builds an oven-shaped nest of grass but belongs to the unrelated wood warbler group.

Ovenbirds are so called because of the rounded shape of the true ovenbird's clay nest, which resembles an earthen oven.

Wide range of habits

Ovenbirds are shy and tend to be overlooked because of their unassuming plumage. The rufous horneros are among the most conspicuous members of the ovenbird family because their domed clay nests are a feature of the open country in Argentina and neighboring states. The nests are often visible on trees, fence posts and sometimes on the eaves of houses. The habits of rufous horneros are rather like those of larks and thrushes that feed on the ground. The miners, genus *Geositta*, and earth-creepers, genus *Upucerthia*, are also terrestrial, and some species prefer to run rather than fly when disturbed. The cinclodes, genus *Cinclodes*, can often be found along watercourses, continually flicking their tails in the manner of their aquatic counterparts, the wagtails. Some cinclodes are even found along the shores of Chile, where they feed among the floating kelp. This makes them the only passerine (perching, singing bird) to have taken on a partially marine way of life. The stream-creeper, genus *Lochimas*, also lives by streams, favoring those tainted with sewage effluent.

Other members of the family are marsh birds, and many live in woods and forests. The treerunners, genus *Margaronis*, have stiff tails, which they use as props when climbing tree trunks, as do nuthatches and woodpeckers. The foliage-gleaners, genera *Philydor* and *Automolus* among others, search for insects among the leaves, as warblers do. While on the ground, among dense undergrowth, the leafscrapers, genus *Sclerurus*, search for insects among fallen leaves, which they toss in the air in the manner of thrushes.

Similar diet

Despite their great range of habits, most ovenbirds feed primarily on insects. The kind of insects that they take no doubt depends on the

One rufous hornero nest may consist of up to 2,500 lumps of clay. Beyond the narrow entrance is a grass-lined chamber about 8 inches (20 cm) wide.

places where the ovenbirds live. Those cinclodes that forage among kelp beds feed on small crustaceans and other marine animals, and a few species feed on seeds.

Ovenbirds have varied songs, ranging from the ringing cries of the miners to the harsh jaylike screams of the brown cachalote, *Pseudoseisura lophotes*, and the tuneful duets of the rufous horneros. Many ovenbirds either build solid nests out of clay or plants or nest in burrows. The birds' breeding habits are not yet fully known. The eggs, which usually number three to five, are generally white. Incubation lasts in studied species for 15–20 days, and the chicks normally fledge in 13–18 days.

Structure of the nest

Studying the nesting habits of ovenbirds is problematic, because the contents of many nests cannot be examined without destroying them. The rufous hornero makes its nest of clay strengthened with grass, building up the walls until they are 1½ inches (3.8 cm) thick. The nest is built in the winter when the rains make the clay soft enough to be malleable. The pair first build a cup and then continue adding to the walls until the nest is roofed in. The nesting chamber, which is lined with grass, is reached by a curved corridor. The nest can be massive, weighing as much as 7½ pounds (3.5 kg). It is made of mud, pieces of stick, straw and animal hairs. Three to five white eggs are laid on a bed of soft dry grass in a large chamber, reached via an antechamber.

RUFOUS HORNERO

CLASS **Aves**

ORDER **Passeriformes**

FAMILY **Furnariidae**

GENUS AND SPECIES ***Furnarius rufus***

ALTERNATIVE NAMES
Ovenbird. There are many other species of ovenbirds, some with alternative names.

WEIGHT
About 1 oz. (30 g)

LENGTH
Head to tail: 7–8 in. (18–20 cm)

DISTINCTIVE FEATURES
Rufous-brown upperparts and tail; paler buff-brown underparts

DIET
Insects

BREEDING
Age at first breeding: 1 year; breeding season: variable; number of eggs: 3 to 5; incubation period: 15–20 days; fledging period: 15–18 days; breeding interval: 1 year

LIFE SPAN
Not known

HABITAT
Open habitats below 8,200 ft. (2,500 m)

DISTRIBUTION
Northeast Brazil west to Bolivia and south to central Argentina

STATUS
Common

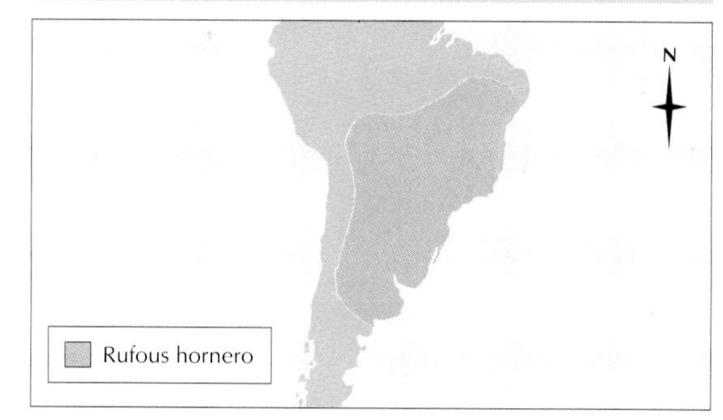

The birds are monogamous and share the duties of incubation. The nest is sufficiently strong to last 2–3 years before being washed away by the rains, but the ovenbirds build new nests each year. After the nests have been abandoned, they are taken over by hornets, wasps and birds such as swallows and cowbirds.

OWLET MOTH

In all there are 25,000 species of owlet moths worldwide belonging to the family Noctuidae. These include the setaceous Hebrew character (*Xestia c-nigrum*), the true lover's knot (*Lycophotia varia*), the beautiful yellow underwing (*Anarta myrtilli*) and Mother Shipton (*Callistege mi*). Most species are nocturnal.

Not surprisingly, there is great variation within this family, one of the largest and most important of the moth families. Owlet moths are found worldwide in an enormous array of habitats including woods, mountains, grassland, agricultural crops, marshland and tropical forest.

Owlet moths range in size from the giant agrippa, *Thysania agrippina*, with its 1-foot (30-cm) wingspan, to the smaller species with wingspans from as small as ⅗ inch (1.5 cm) across. Most owlet moths are dull colored, often gray or brown, and are difficult to see when they are resting on a leaf or tree trunk. Many have patterning, called disruptive coloring, that helps to break up the outline of the body. Others, especially those living in the Tropics, are actually quite brightly colored, while some of the drab owlets have brilliantly colored hind wings, which are hidden under the forewings when the moths are at rest. When these species are disturbed, the wings are spread and the colors suddenly appear. This is called flash coloration and is presumed to be a means of defense, startling a predator sufficiently to allow the moth to escape. Owlet moths that have colored rear wings are often called underwings.

A particular feature of owlet moths and of some moths in other families are the "ears" or tympanal organs situated on the thorax. Although these are hearing organs, their mechanism is different from that of vertebrate ears. The tympanal organs, one on each side of the body, consist of a cavity that is covered by a stretched membrane known as the tympanum.

Night fliers

Most noctuid moths are nocturnal, hiding during the day in crevices or resting on the bark of tree trunks. Here they are able to remain unseen because of their camouflage. A large number of species are, however, active by day and can be seen on flowers in the company of butterflies and bees. Some of these owlets live in Arctic regions, where there is continuous daylight during the summer. The night fliers come out at dusk when, like their daytime relatives, they can be seen feeding at flowers. Owlet moths have a long proboscis or tongue with which they suck nectar from flowers or sip fruit juices and sap.

A few owlet moths migrate, often in vast numbers. For example, the cotton moth, *Alabama argillacea*, of North America, a serious pest of cotton, migrates north in the autumn, coming to rest 1,000 miles (1,600 km) or more north of the Cotton Belt, where it dies without breeding. It is so abundant in the warmer cotton-growing areas, however, that each year there is a surplus that flies north. The silver Y, *Autograhpa gamma*, named after the patterns on its wings, flies from Africa, across the Mediterranean and into Europe during the spring. Unlike the cotton moth, the silver Y breeds in the northern parts of its range.

Say it with scent

For some years it has been known that moths of many species use scent and various other behaviors to attract one sex to the other. The sensitivity of the scent organs in the antennae is remarkable, and a single female can attract males from considerable distances. Male owlets can often be distinguished by their threadlike antennae, on which there are many organs of smell.

Most species of owlet moths, such as this angle shades moth, are patterned in shades of brown or gray. This helps to break up the outline of the body, making the insect hard to see when it is at rest.

A beautiful yellow underwing, one of more than 25,000 species of owlet moths. The family is found worldwide.

The scent is disseminated by structures called brushes. Owlet moths have a pair of brushes that lie in pockets on the back of the abdomen. When at rest, the lips of the pockets are tightly closed so that the scent is kept in. Then, when a female owlet moth is ready to mate, the brushes are lifted out of the pockets and the scent is blown away by the wind. In the angle shades moth, *Phlogophora meticulosa*, for example, mating takes place at dawn. The males remain at rest until the females expose their brushes. They then fly upwind to the females and on reaching them raise their own brushes for a second or two before proceeding to mate.

Caterpillars feed on crops

The female moth lays the eggs singly or in groups on or near the particular plants on which the caterpillars will feed. Sometimes the eggs are laid in the soil. One female is able to produce many eggs, perhaps thousands, depending on species.

Some owlet caterpillars, such as the bollworms (discussed in greater detail elsewhere), are pests on some cultivated crops. The caterpillar pests are sometimes called cutworms or army worms. The latter name is given when vast masses of caterpillars exhaust one supply of food and migrate in search of another. The caterpillars usually feed at night, lying up by day in crevices or in bunches of leaves that are fastened together with silk. The pupae are usually formed in crevices in the ground.

Ultrasonic tactics

Bats are the main predators of owlet moths and other moths that fly by night. It is now well known that bats search for and track down their prey by echolocation or sonar. The sonar system has been shown to be very sensitive, and it would seem that bats ought to have no difficulty in catching their prey. Experiments have, however, shown that owlet moths and other moths, as well as lacewings, are able to take avoiding action. The tympanic organs of owlet moths are sensitive to the ultrasonic squeaks of some species of bats and evasive action may be taken by the moths, such as flying away from the bat or dropping to the ground.

OWLET MOTHS

PHYLUM **Arthropoda**

CLASS **Insecta**

ORDER **Lepidoptera**

FAMILY **Noctuidae**

GENUS AND SPECIES **About 25,000 species in many genera, including cotton moth, *Alabama argillacea*; beautiful yellow underwing, *Anarta myrtilli*; large yellow underwing, *Noctua pronuba*; and giant agrippa, *Thysania agrippina***

ALTERNATIVE NAMES
Noctuid moth; cutworm, army worm (larva of certain pest species)

LENGTH
Wingspan: ⅔–12 in. (1.5–30 cm)

DISTINCTIVE FEATURES
Broad hind wings; threadlike antennae (male only); most species dull gray or brown, some species with brightly colored underwings; tropical species often have bright colored forewings

DIET
Adult: flower nectar, sap and fruit juices. Larva: plants; some species cannibalistic under extreme circumstances.

BREEDING
Number of eggs: may be thousands; larval period: from a few months to 1 year

LIFE SPAN
Depends on species and environment

HABITAT
Most habitats including woods, grassland, mountains, marshes and cultivated crops

DISTRIBUTION
Worldwide

STATUS
Varies; many species common

OWLET-NIGHTJAR

Owlet-nightjars are perfectly camouflaged when they perch on branches or among the leaf litter on the forest floor. The mountain owlet-nightjar, A. albertisii, *is pictured.*

THERE ARE NINE SPECIES of owlet-nightjars in the family Aegothelidae, part of the order Caprimulgiformes, which also includes the nightjars and frogmouths. The owlet-nightjars are neither owls nor nightjars or frogmouths, but they do share similarities with these birds. An owlet-nightjar resembles a small owl, having a rounded head and large, blackish brown eyes. It has a weaker, shorter bill than is usual in other members of the family, surrounded by prominent bristles. The feet are weak and pink in color.

When it is perched, an owlet-nightjar's posture is almost as upright as that of an owl, but the tail is fairly long, with rounded feathers. The smallest owlet-nightjar species is about 6 inches (15 cm) long and the largest is no more than 12 inches (30 cm). The plumage is dark gray brown on the back and lighter on the front, with patterns that resemble those of an owl's plumage. One notable feature is the tufts of hairy feathers, with separated barbs, that stick up from the forehead and chin. The nine species of owlet-nightjars are very similar in appearance. They are found only in Australia, New Guinea, the Moluccas and New Caledonia. The most widely distributed species is the Australian owlet-nightjar, *Aegotheles cristatus*.

Owlet-nightjars call from their hollows during the day, although they are not active during this time. The call is a double churring note or a whistle. If it is disturbed inside a tree, an owlet-nightjar peers out, owl-like, and hisses at the intruder. If it is disturbed further, the bird flies directly to another branch, where it perches facing the source of the disturbance, even though it can turn its head through 180°, allowing it a wide scope of vision.

Direct flight

Owlet-nightjars are nocturnal birds, and are usually seen most often at twilight. They are occasionally visible during the day when the sky is overcast, and sometimes sun themselves at the entrance of their hollow. Their favored habitat is wooded country and they hide during the day in hollow branches or dense foliage, although some species can be found in grassland. Their flight is more direct than that of nightjars and frogmouths, and they perch on branches in a crosswise position rather than lengthwise as other members of the family do.

The diet consists mainly of insects, particularly beetles. The birds sometimes catch their prey on the wing, but more frequently take them

Owlet-nightjars share a number of features with owls, including their large, forward-facing eyes and flexible necks. The Australian owlet-nightjar (above) is the most widespread of the nine species.

from the ground. Thus, their feeding habits are intermediate between the aerial hawking of nightjars and the ground feeding of frogmouths.

Two broods a year

Owlet-nightjars lay their eggs in the same hollow branches that they roost in, though they are also laid in tunnels in banks. Sometimes the hollow is lined with green leaves or fur, but it is more usual for it to be left bare. The eggs, three or four in number, are white, sometimes spotted, and almost spherical, nearly 1 inch (2.5 cm) in diameter. The eggs are laid at intervals of 1–2 days. Incubation is by the female, possibly assisted by the male, for 25–27 days. When they hatch, the chicks have long white down, which is replaced by gray down after about 1 week, and then by juvenile plumage. Both parents feed the young. Fledging takes place between 21–29 days. Juvenile birds may remain with the parents for several months after fledging, but this has not been confirmed.

AUSTRALIAN OWLET-NIGHTJAR

CLASS **Aves**

ORDER **Caprimulgiformes**

FAMILY **Aegothelidae**

GENUS AND SPECIES ***Aegotheles cristatus***

ALTERNATIVE NAMES
Fairy owl; little owl; moth owl; banded goatsucker; crested owlet-nightjar

WEIGHT
1¼–25½ oz. (35–65 g)

LENGTH
Head to tail: 8¼–10 in. (21–25 cm)

DISTINCTIVE FEATURES
Owl-like, rounded head; very large, forward-facing eyes; tufts of bristles around short bill; longish tail; dark grayish brown upperparts with pale spots; pale eyebrows; buff underparts with brown vermiculations

DIET
Insects, particularly beetles

BREEDING
Breeding season: August–December; number of eggs: 3 to 4; incubation period: 25–27 days; fledging period: 21–29 days; breeding interval: 1 year

LIFE SPAN
Not known

HABITAT
Open woodland, shrubland and trees along water courses; roosts and nests in hollow branches or tree trunks

DISTRIBUTION
Australia, Tasmania and southernmost New Guinea

STATUS
Common or fairly common

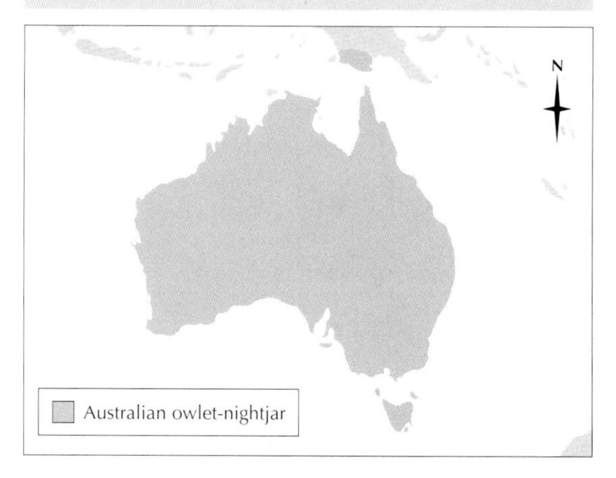

OWLS

Few birds are the subject of so many popular misconceptions as the owls. Although we think of them as night birds, not all owls are truly nocturnal. Only two-thirds of the 134 species of owls hunt at night, and most of the rest are crepuscular, that is, they are active mainly at dusk and dawn. A number of species regularly hunt by day, and some always do so. A second widely held misconception is that owls are birds of prey. Despite having strong, hooked bills, powerful talons and acute senses of sight and hearing, owls are in fact only distantly related to eagles, hawks, buzzards and falcons. Owls are classified in a different order, Strigiformes, and evolved separately from the true birds of prey, or raptors, which belong to the orders Accipitriformes and Falconiformes. Birds of prey are discussed in a separate guidepost article in this encyclopedia.

Although it is a very large owl, the great gray owl preys entirely on small rodents such as lemmings, voles and gophers.

Ancestors and fossil history

Owls are probably descended from the same ancestors as the nightjars, nighthawks, whippoorwills, frogmouths and potoos, which make up the order Caprimulgiformes. Most of the birds in this order are active in the hours of darkness and live in temperate or tropical regions. They have subdued, mottled plumage that provides excellent camouflage in their daytime roosts. However, whereas the Caprimulgiformes are insect eaters, owls hunt a wide range of small animals, especially rodents. Ornithologists believe that the owls and the nightjars and their relatives started evolving into two separate groups about 100 million years ago.

The oldest owl fossils from North America are about 60 million years old, while owls that inhabited what is now western Europe have been identified from fossils 30 to 45 million years old. These prehistoric owls probably looked intermediate between the barn owl, *Tyto alba*, on the one hand, and the North American barred owl, *Strix varia*, on the other.

CLASSIFICATION
CLASS Aves
ORDER Strigiformes
FAMILY Tytonidae: barn owls; Strigidae: typical owls
SUBFAMILY Tytoninae: barn owls; Phodilinae: bay owls; Buboninae: eagle owls and relatives; Striginae: wood, forest or eared owls and relatives
NUMBER OF SPECIES 134

Classification

There are two families of owls: the barn owls, Tytonidae, which contains just 10 species, and the typical owls, Strigidae, which includes the remaining 124 species. While the barn owls are, broadly speaking, very similar to one another, the typical owls differ greatly in size, appearance and habits. The smallest owl, the least pygmy owl, *Glaucidium minutissimum*, is just 4¾–5½ inches (12–14 cm) long and would be dwarfed beside the largest species, the eagle owl, *Bubo bubo*, which grows to 2½ feet (75 cm) long. Like many birds of prey, the female is often larger than the male, a phenomenon known as sexual dimorphism.

Most authorities think that there are around 25 genera of typical owls. However, owl taxonomy is currently the subject of much debate, and zoologists disagree as to the exact number of genera, species and subspecies. Moreover, new kinds of

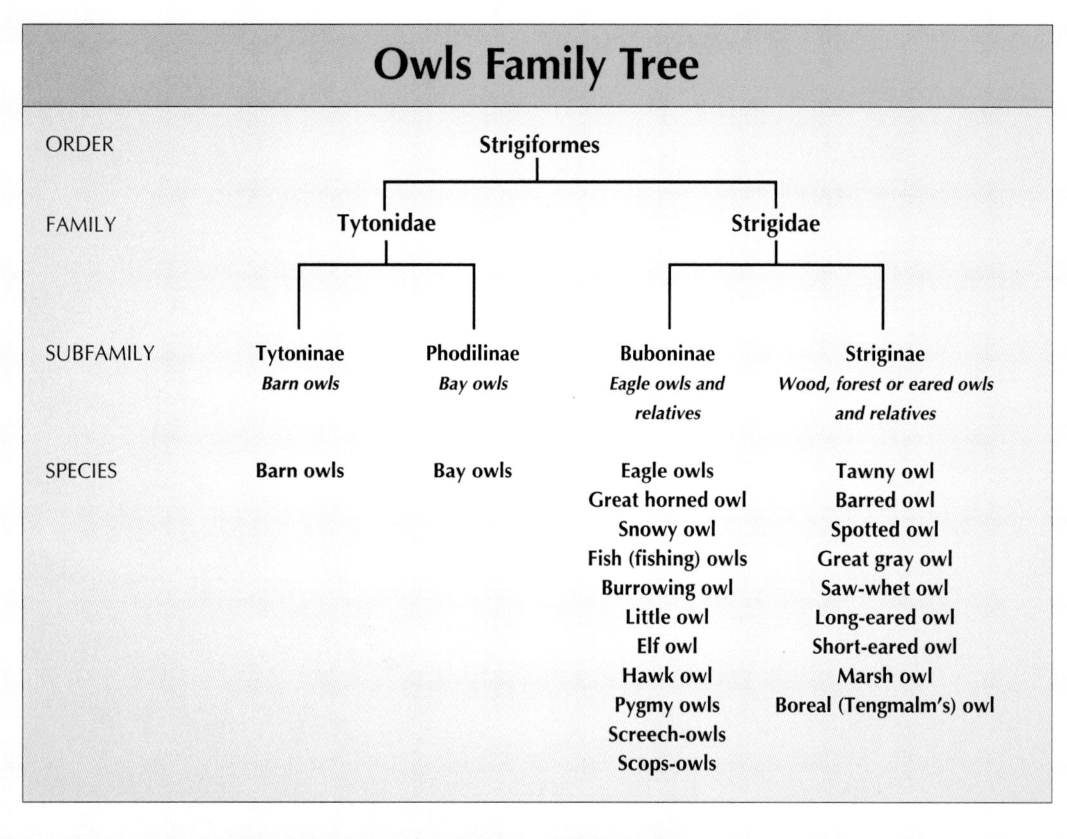

owls are still being discovered and some species remain poorly known. For example, in 1997 the critically endangered forest owlet, *Athene blewitti*, of dry forests in western India, was seen for the first time since 1884. Until its rediscovery, this species had been presumed extinct. A new species of scops-owl was identified in Madagascar as recently as 2000.

Special adaptations

Owls are soft-plumaged birds with short tails and large, rounded heads. They have huge, forward-facing eyes—in common with many nocturnal predators—which are surrounded by a broad facial disc. The eyes are shaped like tapering cylinders to provide the largest possible expanse of retina, maximizing their light-gathering power even further. Owls probably possess the most frontally situated eyes of all birds. This, together with their ability to blink with the upper eyelids, gives owls a semihuman appearance, in which lies much of their appeal to people. Although the eyes themselves are almost immobile, an owl can rotate its whole head to look sideways and even over its shoulder. The long-eared owl, *Asio otus*, is able to rotate its head through at least 270°.

The barn owls have proportionately smaller eyes than the so-called typical species and are easily recognizable by their distinctive heart-shaped facial disc, which is generally more rounded in the typical owls of the family Strigidae. Barn owls also have longer, more slender legs.

The dense, soft plumage of owls makes them look bigger than they actually are. It keeps them warm during their long periods of inactivity during the day, when they perch almost immobile on a suitable perch or in a hole in a tree trunk. The owls' specialized feathers also act as silencers when the birds are in flight, enabling them to swoop on prey in virtual silence. Owls are light in relation to their wing area (in aeronautical terms they have a low wing-loading), which explains why they can fly buoyantly and effortlessly, with relatively little wing flapping.

Many owls nest in holes in trees, but the elf owl, an inhabitant of treeless deserts and canyons, uses saguaro cacti instead.

Sight or sound?

At one time it was thought that all owls hunted by sight. Certainly they have very keen sight, perhaps a hundred times as keen as that of humans, but experiments have shown that some owls can catch their prey in total darkness, where it is absolutely impossible to see anything. A detailed examination of an owl's ears shows them to be very well developed, and there are flaps of skin forming outer ears hidden under the feathers. These flaps are not placed symmetrically about the

Owls (little owl, Athene noctua, *above) have specialized feathers that enable them to swoop on prey in virtual silence.*

head, which means that sound coming to one ear follows a slightly different path from that going to the other ear. Thus a sound is picked up by one ear slightly before or after the other. It is this fractional difference in timing that enables an owl to judge the position of its prey.

Usually owls locate prey by waiting until they hear the rustling of leaves or other vegetation, or by listening for the high-frequency squeals made by rodents. The great gray owl, *Strix nebulosa*, has the remarkable skill of being able to pinpoint from the air invisible small mammals that are moving along tunnels beneath a blanket of snow. The owl leaves its perch, swoops low over the ground and then crashes down through the snow immediately above the lemming or vole.

Habitat

Owls live in a wide variety of habitats, including rain forest, temperate woodland, open grassland, tundra, marshes, scrub and semidesert. However, only a few species inhabit areas where trees cannot grow; they include the snowy owl, *Nyctea scandiaca*, which is confined to the open tundra of North America, Greenland, Scandinavia and Siberia, and the elf owl, *Micrathene whitneyi*, which occurs in the treeless deserts and canyons of Texas and Central America. The Eurasian tawny owl (*Strix aluco*), the two common North American screech owls (*Otus asio* and *O. kennicotti*) and several other species are often found in suburban parks and gardens. The seven species of fish or fishing owls (genera *Scotopelia* and *Ketupa*) frequent swamps or riverside woodland, as does the barred owl. Most of the wooded islands in the Florida Everglades have a resident pair of barred owls.

Sometimes habitat preference is virtually the only distinguishing feature between two or more similar species of owls. For example, the Torotoroka scops-owl, *Otus madagascariensis*, lives in relatively dry forests in the west and center of Madagascar, while the almost identical rain forest scops-owl, *O. rutilus*, occurs in humid rain forests in the east of the island

Camouflaged plumage

Owls generally have cryptic (well-camouflaged) plumage. Woodland owls tend to be brown or gray in basic coloration, owls living in scrub and more open habitats are typically paler, and those inhabiting arid country and desert are usually sandy colored. Hume's desert owl, *Strix butleri*, a rare and highly localized species found in the Middle East, is very pale buff gray. Its close relative, the far more common and widespread tawny owl, has a complex brown plumage with an intricate pattern of dark and pale markings, which looks exactly like the tree bark of its woodland home. The snowy owl has a mainly pure white plumage with variable amounts

A breeding pair of the critically endangered Sokoke scops-owl,* Otus ireneae, *restricted to the Sokoke forest in Kenya.

of black spots and chevrons. Snowy owls blend in perfectly with the snow-covered landscapes of the Arctic during the long northern winter.

Vocal communication

Throughout the world owls have evolved an extraordinary repertoire of calls, including hoots, hisses, shrieks, screeches, barks and whistles. These calls are one of the main nocturnal animal sounds in temperate regions, whereas in the Tropics owl calls are just part of a richly varied chorus of noises made by, among others, insects, frogs, toads and small mammals. The calls of owls carry far in the night air and serve to announce the presence of occupied feeding and breeding territories. For this reason, in temperate regions such as North America owls are most often heard in late winter and early spring, when they are establishing territories before the start of the breeding season.

Some owls look very similar even to experienced human observers, but the owls can tell each other apart by their voices. The differences in their DNA prove that they do not interbreed and so are separate species. The eastern screech owl, *Otus asio*, found in the eastern half of the United States, has two typical calls: a long trill and a series of quavering whistles, descending in pitch. The western screech owl, *O. kennicotti*, which lives to the west of the Rockies, makes two, quite different calls. One is a short trill immediately followed by a longer trill, which is performed as a duet by the male and female. The second type of call is a series of brief whistles that gradually accelerates in tempo.

Breeding

Owls are monogamous, nesting in separate pairs. They often pair for life, mating with the same partner in successive seasons. Ornithologists have found that it is possible to recognize individual owls by their voices alone, and it is almost certain that the owls can do the same.

In general owls do not build their own nests. Most species raise their families inside holes in tree trunks, or use the old nests of other birds such as crows or raptors (birds of prey). Some owls, including the snowy owl and the short-eared owl, *Asio flammeus*, lay their eggs in a simple hollow on the ground. Other species nest in more unusual places. Eagle and Hume's desert owls nest on cliff faces, sometimes in small caves. The tall columns of saguaro cacti make ideal nesting sites for the elf owl, which is the smallest North American owl, measuring just 6 inches (15 cm) in length. Burrowing owls, *Athene cunicularia*, take over the abandoned burrows of mammals such as prairie dogs, groundhogs, armadillos and badgers. If they are threatened by a predator, young burrowing owls confuse it by making a loud hissing that resembles a rattlesnake.

Owls lay from 1 to 14 eggs, the clutch varying in size between species, although it is largely dependent on the food supply. For example, in years when lemmings are abundant, the snowy owl may lay 10 to 14 eggs, and in years of food scarcity it may have clutches of just three or four. Owls lay their eggs several days apart, and incubation starts with the first egg laid, resulting in marked differences in the size of the chicks. The youngest chicks tend to starve in all but the best years, when food is plentiful. In most species of owls only the female incubates, while the male forages and brings food to her. The incubation period varies from 26–36 days, according to species, and the fledging period is 4–8 weeks (being longer in the largest species). Compared to many other birds, owls are relatively long-lived, and there are several records of owls living for more than 20 or even 30 years in captivity.

Problems of study

Owls are harder to study than most other birds. They are more often heard than seen, and the subdued plumage of many species provides them with superb camouflage. Owls also have acute hearing and can drift silently away long before a human has approached close enough to have a chance of seeing them. In addition, owls often occupy large territories and, like birds of prey, usually live at comparatively low population densities.

One way to study owls is to play recordings of their calls. An owl will often respond to a taped call of its species. It gradually will move nearer to the source of the sound, calling at regular intervals, to investigate the unwelcome "intruder." When the owl has perched nearby it can be observed by flashlight. However, great care needs to be taken by researchers using this study technique so as not to disturb the owls.

Another method of study, which causes no disturbance at all to the owls, is to collect and examine the pellets that they regurgitate on a daily basis. The pellets contain the indigestible remains of the owls' meals, including bones, fur and insect wing cases, and piles of them gather on the ground underneath the birds' favorite roosting spots. It is from their pellets that the diets of various owl species can be identified. For example, a large number of barn owl pellets were once analyzed in Poland. The remains of nearly 16,000 vertebrates were found and identified. Of these, 95 percent were small mammals, 4 percent birds and the remainder, amphibians. In the United States, the staple prey items of the barn owl are mice, moles, shrews and cotton rats.

Some species of owls will use nest boxes, enabling ornithologists to study their breeding behavior and measure and band the chicks. In Scandinavia owl conservation programs have been so successful that in certain pine forests almost all of the Eurasian pygmy owls, *Glaucidium passerinum*, and boreal owls, *Aegolius funereus*, now use nest boxes rather than natural holes in hollow trees.

Relationship with humans

References to owls exist throughout human culture, from cave paintings, the Bible and ancient Greek legends to Shakespeare's plays and children's nursery rhymes. It is clear that owls leave a powerful impression on us. Despite this popularity, however, a growing number of owl species are threatened with extinction. The World Conservation Union (I.U.C.N.) lists four species as critically endangered, five species as endangered and a further 17 species as vulnerable. In most cases the population declines are due to habitat loss, especially deforestation and the conversion of grassland to intensively farmed agricultural land. This is a particular problem with species that have restricted ranges, such as the Indian forest owlet, the Seychelles scops-owl (*Otus insularis*) and the Grand Comoro scops-owl (*O. pauliani*). However, even globally widespread species such as the barn owl have suffered dramatic declines in parts of their range. Barn owls need a plentiful supply of prey and suitable nesting sites, but rodents, disused buildings and dead or hollow trees have all become much scarcer as a result of modern farming techniques.

Eagle owls are powerful enough to tackle prey such as grouse, ducks and young deer, as well as smaller species of owls.

For particular species see:
• BARN OWL • BURROWING OWL • FISH OWL
• SNOWY OWL • TAWNY OWL

OXPECKER

Oxpeckers enjoy symbiotic relationships with a wide range of African mammals, including the sable antelope,* Hippotragus niger *(above).

OXPECKERS, ALSO KNOWN as tick birds, have a highly distinctive natural history, because they spend most of their time associating with large mammals. They perch on the backs of African game, feeding on the ticks, blood-sucking flies and tissues of their hosts. Rhinoceroses, giraffes, hippopotamuses and eland are favorite big mammalian hosts, but oxpeckers are often seen cleaning smaller ones, such as impalas. The association is at least partly of mutual benefit to both host and bird: the oxpeckers feed, while the mammals have their pests removed and are warned of danger by the alarm calls of the birds. The oxpeckers rely on the relationship and much of their lifestyle is linked to it. Even their claws are adapted to their behavior, being curved and very sharp, to clutch the hide of the host. Their tails are stiff, like those of woodpeckers, to give support on a vertical surface, such as the flanks of a large animal.

The two species of oxpeckers are confined to Africa and are very similar in appearance. They are close relatives of starlings and are about 8 inches (20 cm) long, with short, flattened bills and longish tails. The plumage is almost uniform brown. The red-billed oxpecker, *Buphagus erythrorhynchus*, has a completely red bill, but the yellow-billed oxpecker, *B. africanus*, has a yellow bill with a red tip. In flight it can be distinguished by its paler rump.

The yellow-billed oxpecker ranges from Senegal in the west to Ethiopia in the east and Natal in the south; the red-billed oxpecker is found only on the eastern side of the African continent. The ranges of the two overlap, and at times both can be seen perching on the same animal. It is most unusual for two such closely related animals to live in the same habitat and, apparently, to have the same feeding habits.

Becoming rarer

There may be a dozen oxpeckers on one animal, perched in a line on its head or along its back, or running around on its body. Their sharp claws enable them to cling firmly while their host is galloping at full speed, or to clamber under the belly and up and down the legs with the ease of a woodpecker or nuthatch on a tree trunk. Although the hosts rarely object to oxpeckers, they do sometimes try to drive them off by a flick of the tail or by rolling on the ground, particularly if they have open sores or wounds.

At one time oxpeckers could be found almost wherever there was wild game or cattle, but their range has diminished considerably as the game has been killed and as farmers have begun the practice of dipping domestic stock in disinfectant. The poisonous dips kill ticks, one of the oxpeckers' main sources of food, and probably also kill oxpeckers that eat the poisoned ticks. Oxpeckers have disappeared entirely from some parts of their range, although generally they remain common. Some farmers, both African and European, have not regretted the disappearance of the oxpeckers because of the damage they do to hides by enlarging wounds and aggravating sores. It is also claimed, without evidence, that oxpeckers spread diseases such as rinderpest. In fact, they are more likely to be beneficial because they eat the blood-sucking flies that are known to be carriers of disease. Some cattle-farming tribes, such as the Fulani of Gambia and the Masai of Kenya, realizing this, encourage oxpeckers.

Blood drinkers

Although oxpeckers feed mainly on insects plucked from the host's back or caught as they fly near the host, they occasionally also feed on carrion. Ticks and flies make up the greatest number of prey taken but some lice, as well as mites, are also eaten. The birds cut pieces of flesh out of sores and wounds, which are also kept clear of maggots. Oxpeckers remove ticks and pieces of skin by a scissoring action of the flattened bill, which is laid sideways on the skin. They drink the moisture from around their hosts' eyes as well as blood. The blood the ticks have gorged seems to be more important to the oxpeckers than the tissues of the ticks. In a series of experiments, the British ornithologist Derek Goodwin demonstrated the oxpeckers' appetite for blood by presenting a cut finger to a captive oxpecker, which immediately pecked at it.

RED-BILLED OXPECKER

CLASS **Aves**

ORDER **Passeriformes**

FAMILY **Sturnidae**

GENUS AND SPECIES ***Buphagus erythrorhynchus***

WEIGHT
About 1¾ oz. (51 g)

LENGTH
Head to tail: 8–8¾ in. (20–22 cm)

DISTINCTIVE FEATURES
Short, flattened, red bill; orange iris; strong claws; stiff, woodpecker-like tail; brown upperparts, throat and breast; paler rump; creamy buff underparts

DIET
Mainly ticks, horseflies, lice and mites, taken from hide of host mammals; also wound tissue, dry skin flakes and blood of hosts

BREEDING
Age at first breeding: 1 year or less; breeding season: varies according to rains and movements of host mammals; number of eggs: usually 2 to 3; incubation period: 12–13 days; fledging period: 26–30 days; breeding interval: up to 3 broods per year

LIFE SPAN
Not known

HABITAT
Savanna

DISTRIBUTION
Sub-Saharan Africa, mainly in the east

STATUS
Common

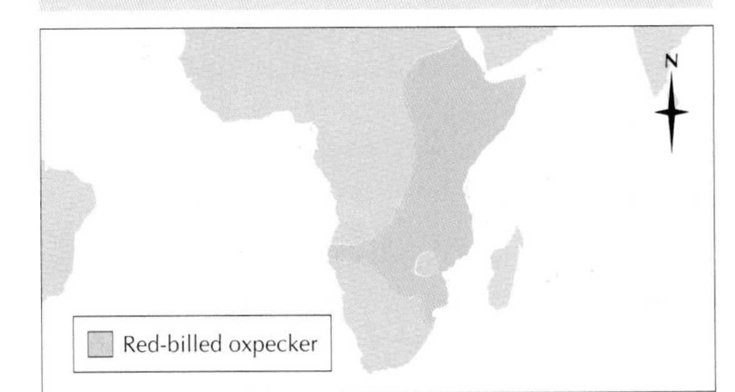

The oxpeckers' hosts provide more than food. The birds regularly bask on the warm backs of the big animals and also mate there. Breeding occurs during the rainy season. The nest is built in a hole in a tree or in a rock, in the eaves of a house or among thatch. It is made from dry grasses and is lined with a pad of hair plucked from the mammals' backs. The clutch usually consists of two or three eggs, though up to five are not rare. Red-billed oxpeckers may have three broods in a season. Incubation of the eggs is by both sexes, though only the female incubates them at night. The nestlings are fed by both parents and up to three helpers.

Oxpeckers' favorites

Zoologists believe that oxpeckers do not choose their hosts at random, nor do they merely settle on the animals with the most ticks. There is evidence that they keep returning to a particular animal in a herd, sometimes even when it has died. Some species of big game are preferred as hosts, but the preference varies from place to place. Oxpeckers are rarely tolerated by elephants, but rhinoceroses are common hosts.

Oxpeckers are very unpopular with hunters because their alarm calls alert the animals with which they associate. This is apparently not just a matter of the hosts learning to associate the oxpeckers' calls with danger. There have been instances of oxpeckers trying to alert their hosts by flying around their heads and calling to them.

Its bright red bill best distinguishes the red-billed oxpecker from its nearly identical yellow-billed relative.

OYSTER

THE EUROPEAN FLAT OYSTER, *Ostrea edulis*, is one of the true oysters of the family Ostreidae. Its two valves, or half-shells, are rough and irregular in outline. There are large local differences in appearance and with experience it is possible to tell in which bed a particular oyster lived. The two valves are unlike each other, the right one being flat and the left convex. They are hinged in the pointed region of the beaks, held together by a triangular elastic ligament. Oysters do not have hinge teeth, as many bivalves do. On each valve a series of wavy ridges centers around the beak, marking former positions of the margin. Rough ridges radiate from the beak of the left valve, while the other valve bears horny scales that are less rigid than the rest of the shell, allowing the valves to make firm contact around the edges when they are pulled together.

Oysters are notable for the soft, porous, chalky masses that are laid down within the substance of the shell. The convex valves of older oysters often contain chambers filled with sea water smelling of hydrogen sulfide, caused by the putrefaction of organic matter. These chambers are the result of the shrinkage of the surface of the mantle, the tissue that lines and secretes the inside of the shell, during the oyster's life. Oysters live for 7–10 years, perhaps longer.

The mantle surrounds the oyster's stomach, heart, nervous system and reproductive organs. When the valves are open, the edges of the mantle are visible. This thickened mantle edge has short sensory tentacles and a muscular fold that controls the flow of water. If the shell opens further, the large central adductor muscle that closes the valves against the pull of the hinge ligament becomes visible. Arranged more than halfway around this and the general body mass are two pairs of crescentic gills.

True oysters such as the European flat oyster have been cultivated as a food for thousands of years. Great mounds of shells in coastal regions all over the world testify to the importance of oysters in the diets of many prehistoric communities. The Romans regularly feasted on oysters, importing them from British oyster beds. During much of the 18th and 19th centuries British oyster beds were the most productive in Europe.

The oyster family includes two other genera, *Pycnodonta* and *Crassostrea* (formerly *Gryphaea*). Among the latter are the American, Portuguese and Japanese oysters, which are consumed more widely than the sweeter European flat oyster. The Portuguese oyster, *C. angulata*, introduced into France in 1868, was relayed to beds on the east coast of Britain during the mid-20th century, but it seldom breeds there now. *Crassostrea* is easily distinguished from *Ostrea* because its shell is elongated rather than round, its left valve is more deeply convex and the muscle scars inside are deep purple.

Apart from these true oysters, there are other bivalves that bear the name oyster. These include the tropical pearl oysters (family Pteriidae), which are closer to the mussels and are, like them, attached by byssus threads; the thorny oysters (genus *Spondylus*); and the saddle oysters (genus *Anomia*). Saddle oysters live attached to rocks by thick calcified byssuses, tufts of long, tough threads, that pass straight through a notch in the lower valve.

An oyster's shell is lined with tissue known as the mantle. Light-sensitive sensors in the mantle enable the oyster to sense predators and close its shell in defense.

Several ways of feeding

The European flat oyster ranges down the Atlantic coast of Europe from Norway (latitude 65° N) as far south as

EUROPEAN FLAT OYSTER

PHYLUM **Mollusca**

CLASS **Bivalvia**

ORDER **Eulamellibranchia**

FAMILY **Ostreidae**

GENUS AND SPECIES ***Ostrea edulis***

LENGTH
Up to 4 in. (10 cm)

DISTINCTIVE FEATURES
Thick, irregularly shaped shell; flat right valve (half-shell), convex left valve; color is off-white, yellowish or cream; light brown or bluish concentric bands on right valve; inner surface of both valves pearly white

DIET
Filter feeder

BREEDING
Sexes separate, but oyster may change sex many times during lifetime. Number of eggs: about 1 million; larval period: 1–2½ weeks.

LIFE SPAN
Usually 7–10 years; perhaps up to 20 years

HABITAT
From low-water mark down to about 260 ft. (80 m); forms thick beds on hard substrates in creeks, estuaries and sheltered water

DISTRIBUTION
Coastal waters from Norway south to Mediterranean; also in Black Sea

STATUS
Common

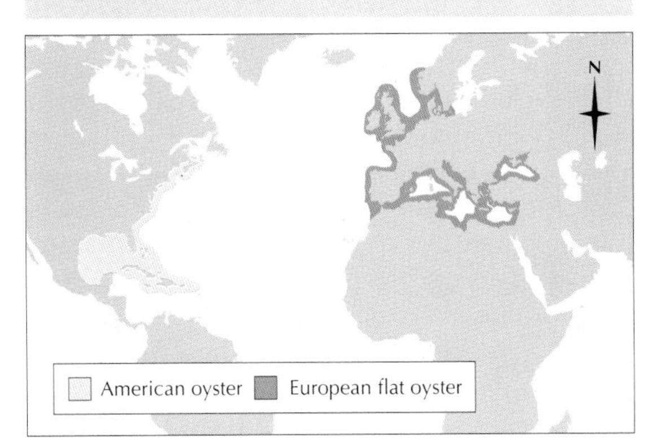

Morocco and also in the Mediterranean and Black Seas. It first settles as a ciliated sphere about 1 millimeter in size, at which stage it is known as spat. The oyster attaches itself to the sea bottom by its convex left valve, although later it may become detached and turned over.

Oysters are sedentary bottom-dwellers, so become encrusted with a variety of aquatic animals and plants, which help to camouflage them.

The adult oyster then stays in one place, feeding by filtering small particles from the water. By beating the tiny hairlike cilia on its complex, latticelike gills, it draws a current of water in at a rate of perhaps 2–3 gallons (9–13.5 l) per hour. The food particles, caught in mucous strings on these ciliated sieves, are wafted either to the bases or to the free edges of the gills and then forward to the mouth via the labial palps that sit on either side of it. The palps both taste and sort food particles. Once in the digestive tract, the particles continue to be propelled by cilia to the elaborate ciliated stomach. It is in the stomachs of bivalves and a few snails that perhaps the only truly rotating structure in the whole animal kingdom occurs. This is the crystalline style, a rod of solid digestive enzymes, rotated by cilia in the sac that secretes it. The style dissolves at the tip, where it rubs against a piece of cuticle, called the gastric shield, and helps to reel in the byssus threads.

In addition to the digestion that takes place in the cavity of the gut, cells lining the gut take in particles and digest them in their cytoplasm. Some of the particles are engulfed by amoeba-like blood cells that come out into the gut cavity through the stomach walls and then migrate back with the food particles trapped in them.

At times the oyster's mantle cavity becomes in danger of becoming clogged with sediment. If this occurs the valves are clapped shut, which suddenly expels the water and sediment. This sudden cough contrasts with the sustained closure of the shell in response to external dangers. The adductor muscle is composed of two parts. One part can contract rapidly, while the other contracts more slowly but can remain contracted for long periods without tiring.

Bisexual bivalves

An European flat oyster may change sex many times during its life, a characteristic known as rhythmical hermaphroditism. This is not rare among bivalves, as their reproductive systems are so simple that the change involves little reorganization. Maturing first as a male, the oyster takes some weeks to become a functional female but recovers its maleness within days of discharging the eggs. In the cold waters off Norway, an oyster may change sex once a year, but in warmer waters this may occur many times.

Spawning occurs in summer, when the water temperature exceeds about 60° F (15° C). The eggs pass through the gills against the water current and are fertilized in the mantle cavity by sperms carried in by the feeding current. They are not freed for about another 8 days, when the shell opens wide and closes violently at intervals, expelling clouds of larvae. Up to a million larvae may be incubated at a time. The American oyster, *Crassostrea virginica*, does not incubate its eggs, but can release more than 100 million eggs. The European flat oyster's eggs are about 0.1 millimeter across. As they develop within the parent's shell they change from white to black.

Explosive spawning

When it is released, the young oyster, now known as a veliger larva, has a tiny shell with two adductor muscles, a ciliated tuft, or velum, for swimming and feeding, and a foot. For between 1–2½ weeks the larva swims in the plankton, but when it is ready to settle, it protrudes its foot and grips any solid object that it touches. It then starts to crawl until it reaches other oysters of the same species, although if an area proves to be unsuitable, it can swim off again to search for a more favorable environment. Eventually, however, if it survives, the larva sticks itself down by its left valve, using a drop of a cementlike substance from the glands that in other bivalves secrete the byssus threads. At this stage, the oysters have become spat. From this point on the shell grows rapidly and the body changes dramatically. The foot, velum and eyes are lost, together with the anterior adductor muscles. The gill increases in size and the mouth moves through 90° to its adult position.

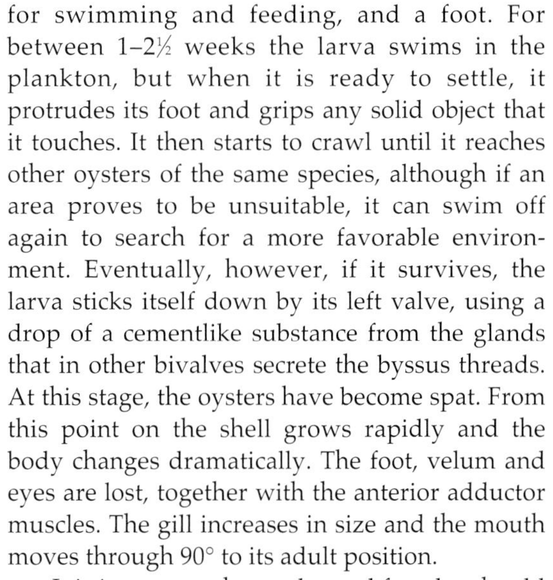

It is important that males and females should spawn at the same time, and to some extent this is aided by the dependence of spawning on temperature. In those members of the genus *Crassostrea*, however, chemical stimulation is also important. The sperms carry a hormonelike substance that stimulates spawning in both sexes and the males are also stimulated to spawn by the presence of eggs and by various organic compounds, including one that is also present in seaweed. In this way, one spawning individual can cause the whole population to release their eggs and spawn.

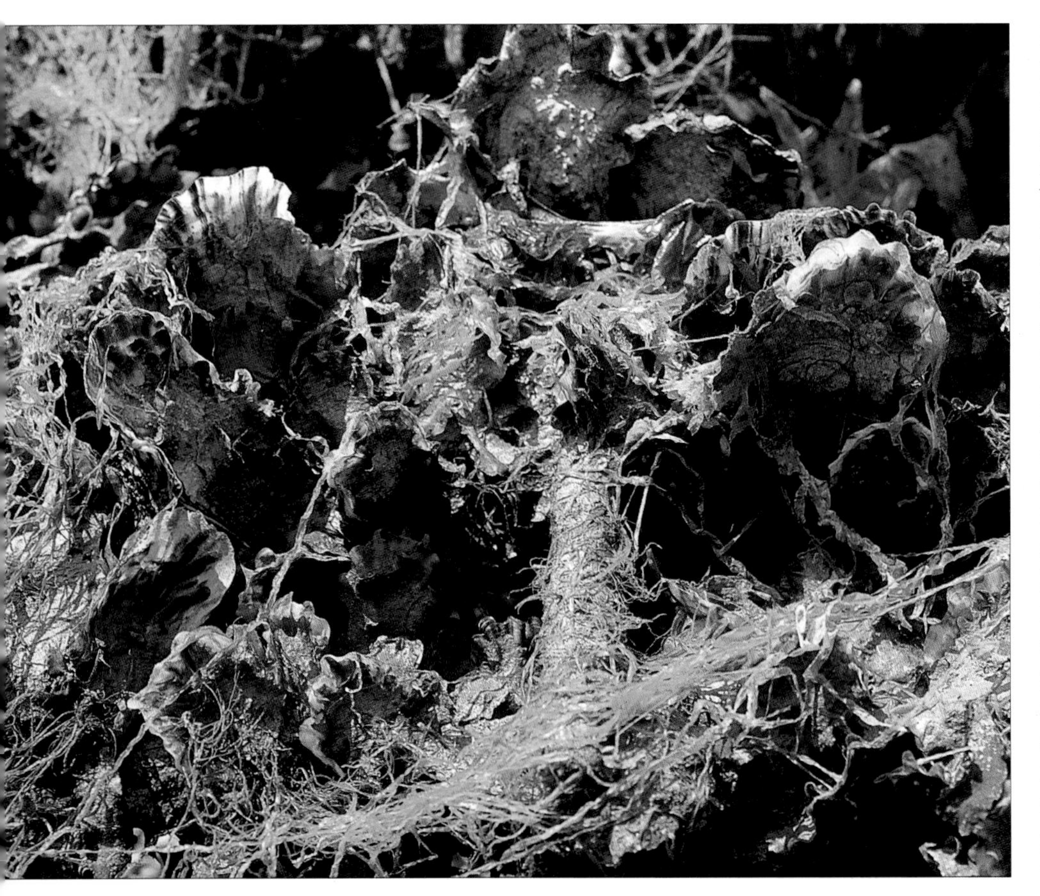

After its larval stage is complete, an oyster attaches itself to rocks or the sea bottom using its left, convex valve. These European flat oysters have been exposed at high tide.

Making pearls

Pearl oysters belong to the family Pteriidae, and mostly belong to the genus *Pinctada.* Pearls are created as a reaction to the introduction of a foreign body, which irritates the oyster's soft inside. The oyster slowly secretes several layers of nacre, or mother-of-pearl, the substance that lines its shell, around the object to make it less irritating, thereby creating a pearl. Pearls produced by edible oysters are lusterless and are of little value. Pearl oysters are widely farmed in Japanese and Australian waters, and in the South Pacific. Most pearls are taken from oysters that are more than 5 years old. Pearl fishing has been practiced in China since before 1000 B.C.E., but has declined since the development of cultured pearls, a process perfected in Japan in the 1890s. Cultured pearls are created by inserting small beads of pure mother-of-pearl into oysters. This is the most effective way to stimulate pearl production, and produces a high quality pearl that consists entirely of nacre.

OYSTERCATCHER

All oystercatchers are coastal birds. They are characterized by pied or black plumage, short legs and a long, straight, bright red bill. The American oystercatcher is pictured.

Oystercatchers are large shorebirds that are found in many parts of the world. Some species have black-and-white plumage, while others are all black. The most widespread species, the Eurasian oystercatcher, breeds in locations from western Europe to eastern Siberia, but can also be found in the Canary Islands, South Africa and Australasia. It is largely black above with white underparts and has a long red bill and pink legs. Another pied oystercatcher is the American oystercatcher, *H. palliatus*, which ranges from New Jersey and California to Argentina and Chile, while a third, the Magellanic oystercatcher, *H. leucopodus*, lives in southern South America. The sooty oystercatcher, *H. fuliginosus*, lives on the coast of Australia, and other black oystercatchers, *H. bachmani* and *H. ater*, live in western North America, southern South America and Australia. The Canary Islands oystercatcher, *H. meadewaldoi*, was also black, but was last seen in 1940.

Coastal birds

Oystercatchers are usually seen on rocky shores or sandy beaches, on mudflats, or in sand dune areas just behind the shore, but they sometimes breed inland. In New Zealand, pied oystercatchers are found by the snow-fed rivers of South Island. Outside the breeding season some species of oystercatchers gather in large flocks, and those that breed in high latitudes migrate to warmer regions in the winter. The Burry Inlet in south Wales, for instance, is the winter home of oystercatchers from Scotland, Iceland, the Faeroes and Norway. The American oystercatcher occurs in smaller concentrations, such as scattered pairs or small flocks. The American black oystercatcher, *H. bachmani,* sometimes associates with other shorebirds such as surfbirds, black turnstones and rock sandpipers. It is sometimes very difficult to see against the black rocks of the west coast of North America.

The all-black or pied plumage and red bill of the oystercatchers are unmistakable, yet, surprisingly, the birds are sometimes difficult to see if they are motionless. They often give away their presence by their loud shrill calls of *kleep-kleep* or a shorter, rapid *kic-kic*. Oystercatchers are wary of newcomers to their area and run rapidly or take flight when approached.

Musselcatchers

The oystercatchers' common name is somewhat misleading. Oysters live below the low tidemark while oystercatchers feed between the tides or on land. A more accurate name would be the old, local name of musselpecker. Mussels, together with limpets, cockles, winkles, crabs and worms, make up a large part of the oystercatchers' diet.

Oystercatchers locate cockles and worms by probing in the sand with their long bills. They also eat aquatic insects, particularly their larvae, as well as plant food and occasionally the eggs of other birds. The precise composition of the diet depends on the animals that live in the oystercatchers' preferred habitat of sandy or rocky shores, mudflats, salt marshes, riversides and damp farmland.

Ornithologists have made detailed studies of the methods by which oystercatchers eat mollusks that are protected by strong shells. Limpets are dealt a sharp blow with the tip of the bill. Small ones are dislodged and large ones are shifted so they can be levered off or holed. The oystercatcher can then insert its bill and tear the strong muscles that hold the limpet down. Two different methods are used for opening bivalve mollusks such as mussels and cockles. If the shellfish is covered with water and its valves, or shells, are agape, the oystercatcher stabs downward and then levers and twists to sever the adductor muscle that closes the valves. These fall open and the flesh is rapidly pecked out. If the shellfish are exposed to the air and firmly closed, the oystercatcher has to smash its way in. Examination of mussel shells that are left over after oystercatcher meals shows that they are regularly smashed on the bottom edge, and tests have shown that this side of the shell is much weaker than the top edge, even in large mussels.

The oystercatcher carries a mussel or cockle to a patch of firm sand, places it with its ventral margin upward and starts to hammer it. If the shell falls over it is righted, and if it sinks it is carried to a firmer patch. On average, five blows of the bill are needed to penetrate a mussel shell, and the bird then inserts its bill to cut the adductor muscle and pry the two halves apart. Cockle shells are not attacked in any particular position, because their shells are weaker than those of mussels. Oystercatchers flip crabs onto their backs and kill them with a stab through the brain. The shell is then pried off and the flesh is cut out with the same scissoring movements that are used for eating other shellfish.

Oystercatchers (Magellanic oystercatcher, below) use their powerful bills to knock mollusks from rocks and to break open their shells.

EURASIAN OYSTERCATCHER

CLASS **Aves**

ORDER **Charadriiformes**

FAMILY **Haematopodidae**

GENUS AND SPECIES ***Haematopus ostralegus***

ALTERNATIVE NAME
Musselpecker (archaic)

WEIGHT
15–23 oz. (430–650 g)

LENGTH
Head to tail: 15¾–17¾ in. (40–45 cm), including 3–3½-in. (8–9-cm) bill

DISTINCTIVE FEATURES
Large shorebird. Black upperparts and head; bright white underparts; long, strong, orange bill; reddish legs.

DIET
Mainly bivalve mollusks, particularly cockles and mussels; also worms, insects and plant matter; occasionally bird eggs

BREEDING
Age at first breeding: 1 year; breeding season: eggs laid April–July; number of eggs: usually 3; incubation period: 24–27 days; fledging period: 28–32 days; breeding interval: 1 year

LIFE SPAN
Usually up to 25 years

HABITAT
Intertidal zone of shorelines; also salt marshes, riversides and damp farmland

DISTRIBUTION
Breeds in coastal areas of Europe (Atlantic and Aegean coasts); also in western Siberia and west-central Asia, and from Kamchatka in eastern Russia south to northeastern China

STATUS
Common

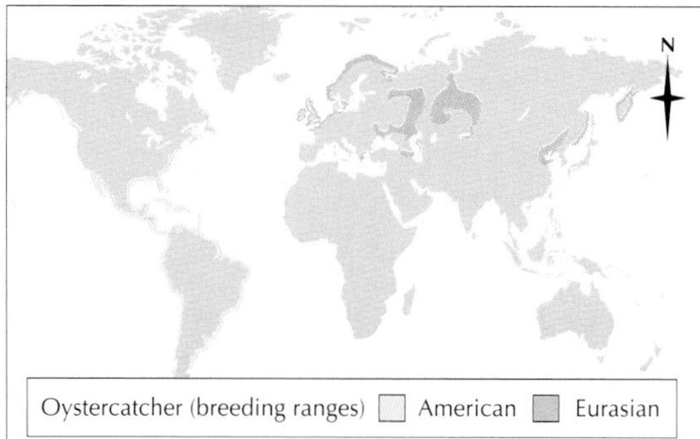

A flock of Eurasian oystercatchers coming in to land at a salt marsh. Oystercatchers are gregarious birds. Outside the breeding season they gather in flocks that may number in the thousands.

In some places, such as the Burry Inlet in south Wales, Eurasian oystercatchers are considered a pest because of the damage they do to cockle beds. On average each oystercatcher eats about one cockle every minute and consumes on average 336 cockles per tide. As flocks number several thousands, they eat many millions of cockles each winter. However, oystercatchers are only one of several enemies of cockles and it is debatable whether they seriously affect the cockle industry. American oystercatchers show many behavioral similarities with their Eurasian counterparts, but tend to be less migratory.

Piping display

Oystercatchers arrive at their breeding grounds in flocks but then split up into pairs. Each pair forms a territory, which it defends against other oystercatchers. Among their several displays there is the quite spectacular piping display in which a group of birds, or sometimes just a pair, run rapidly to and fro with necks outstretched and open bills pointing at the ground. At the same time they utter a piping call that varies from a clear *kleep-kleep* to a quavering trill.

The nest is a shallow depression in shingle, sand or turf, sometimes with no lining but at other times lined with stones, shells or dead plants. There are usually three eggs, yellowish or light brown with spots or streaks of dark brown. Both parents incubate the eggs, which hatch in 24–27 days. The chicks leave the nest after a day or two and are fed by both parents. They fly in about 1 month and are fed by their parents for another 5 weeks.

Family traits

A careful study of the methods Eurasian oystercatchers employ to open mussels was made by M. Norton-Griffiths of Oxford University. He found that some oystercatchers regularly stab open mussels, while others hammer the shells. Furthermore, young oystercatchers develop the same feeding habits as their parents. This is perhaps not so surprising, as the chicks learn to feed on only those animals that their parents bring to them. First the chicks practice pecking empty shells and picking up pieces of flesh left in them, learning the scissoring and cutting movements employed by the adults. Later they take opened shellfish from their parents and remove the flesh on their own. Eventually they open the shells themselves, starting on small ones and graduating to large ones as they become more proficient. Norton-Griffiths never saw a crab-eating chick attack a mussel, and when a mussel-eating chick found a crab it was evidently frightened of it. The differences in feeding habits are so marked that populations of oystercatchers are divided by them. Mussel-eaters mate only with mussel-eaters and cockle-eaters only with cockle-eaters.

PACA

Lowland pacas are nocturnal animals, usually found close to water. They emerge from their burrows at night to forage for seeds, fruits, shoots and insects.

Pacas or spotted cavies are large, stout-bodied, almost tailless rodents. One species, the lowland paca, *Agouti paca*, is found from Mexico to the Amazon Basin, while the other, the mountain paca, *A. takzanowski*, occurs in the Andes from Venezuela south to Peru. Pacas are also sometimes called agoutis, but the true agoutis belong to a different genus.

High- and lowland species

Despite being hunted throughout much of its range, the lowland paca is still numerous in some places. It is stockily built, with a large, broad head, and weighs up to 26½ pounds (12 kg). It has a head and body length of 2⅓ feet (70 cm), with a 1⅕-inch (3-cm) tail. The hindquarters are markedly higher than the shoulders and its brown to black coat is made up of coarse hair. There are a variable number of longitudinal rows of white spots on each side of the rodent's body, and the underparts are white to buff. The lowland paca has large, bulging eyes, moderate-sized ears and the muzzle bears a number of long whiskers. The short legs have four toes on each front foot and five on each hind foot.

The mountain paca is smaller than the lowland species and has a coat of soft, thick fur. It lives in the forests of the Andes, at altitudes of 6,000–10,000 feet (1,830–3,050 m), in cold, humid climates. Its muzzle is more slender than that of the lowland paca, its eyes are less prominent and it is believed to go into semihibernation in the cold season after it has stored fat in its body. Also hunted for its flesh, the mountain paca is classified as being at low risk.

The pacarana or false paca, *Dinomys branicki*, is similar in appearance to the lowland paca but slightly larger. It also lives in the mountain forests of the Andes, but belongs to another family, the Dinomyidae.

Powerful diggers

Lowland pacas usually live in forests close to water. They are strong swimmers and readily use water for escape when danger threatens. They spend the day in burrows, which they dig using all four feet as well as their teeth. Pacas are able to bite through large roots and can also rip through thick plants with their large upper incisors. The burrows are 4–5 feet (1.2–1.5 m) deep, with two or more exits that are often plugged with leaves and vegetation. Mountain pacas also dig burrows in their mountain scrub and heathland habitats.

Nocturnal animals, pacas leave their burrows singly after dusk and make their way to a feeding ground. They eat a variety of stems, roots, leaves and fallen fruits, especially avocados and mangoes. They can be destructive to commercial crops. A peculiarity of pacas is that they do not hold food in their forepaws when feeding, as the great majority of rodents do.

Reinforced cheek bones

The skull of the paca is unusual in having large cheek bones that enclose some of the jaw muscles. This condition is known in only one other rodent, the unrelated African maned rat, *Lophiomys imhausi*. These large cheek bones form the outer walls of capacious cheek pouches and in old males may grow into enormous, blisterlike swellings, giving the head its broad appearance. Another feature of the skull is that part of the cavity enclosed by the large cheek bone is specialized as a resonating chamber.

PACAS

CLASS **Mammalia**

ORDER **Rodentia**

FAMILY **Agoutidae**

GENUS AND SPECIES **Lowland paca, *Agouti paca*; mountain paca, *A. takzanowski***

ALTERNATIVE NAMES
Spotted cavy; laba (Guyana only); agouti

WEIGHT
14–26½ lb. (6.3–12 kg)

LENGTH
Head and body: 2–2⅓ ft. (60–70 cm); shoulder height: 1–1⅓ ft. (30–40 cm)

DISTINCTIVE FEATURES
Stocky body; large, broad head; short legs; hindquarters higher than shoulders; coarse fur, although softer in mountain paca; brown or black upperparts; rows of white spots running along length of back and flanks; white or buff underparts

DIET
Seeds, fruits, shoots, fungi, carrion and insects

BREEDING
Age at first breeding: 1 year; breeding season: early winter, all year in some areas; number of young: 1; gestation period: 118 days; breeding interval: 1 or 2 litters per year

LIFE SPAN
Up to 13 years

HABITAT
Lowland paca: forests, near water. Mountain paca: mountain scrub and heathland.

DISTRIBUTION
Lowland paca: central Mexico into the Amazon Basin. Mountain paca: Andes Mountains from Venezuela south to Peru.

STATUS
Lowland paca: generally common. Mountain paca: at low risk.

In addition to having keen eyesight, the lowland paca's hearing and sense of smell are extremely acute. Although liable to become unduly fat when environmental conditions are favorable, and occasionally with age, pacas are still very fleet of foot.

Single births

Breeding takes place throughout the year in some areas, but in other regions mating occurs in early winter. The gestation period is 118 days and there is normally only a single baby in a litter, although twins have been known. There may be two litters a year. The young is born in the underground burrow and is thought to be little larger than a mouse at birth. It is not weaned for 2–3 months and takes several years to reach full size. Pacas might live for 12 or 13 years in the wild. One in captivity is said to have lived to 16 years of age.

Coveted, tender flesh

Pacas are preyed upon by several carnivores, and their flesh is also prized by the local peoples, who hunt them with dogs. The carcasses fetch high prices in the local markets. Pacas are, however, rugged fighters if necessary and can deliver severe wounds with their teeth. When young or partly grown, pacas are commonly exhibited in zoos, fairs and circuses. Tourists to countries in which they live usually meet pacas under these circumstances, or in a dish on their dinner tables under a variety of innocent or high-sounding names.

Pacas have large cheekbones that enclose some of the jaw muscles. This arrangement is found in only one other rodent, an unrelated African species of rat.

PACIFIC SALMON

In common with the Atlantic salmon, most Pacific salmon return to their natal rivers to spawn. Species such as this chinook may travel upstream for hundreds of miles, leaping waterfalls and other obstacles as they go.

There are 11 species of salmon in the North Pacific, by contrast with the North Atlantic, where there is just one, the Atlantic salmon, *Salmo salar*. Except for the Japanese species, the cherry salmon or masu, *Oncorhynchus masou*, Pacific salmon range from around Kamchatka in Siberia to the West Coast of the United States, occurring as far south as California. Of these, the chinook, *O. tshawytscha*, also known as the tyee, quinnat, king, spring, Sacramento or Columbia River salmon, is the largest, reaching some 5 feet (1.5 m) and weighing up to 135 pounds (61 kg). Other salmon species are much smaller, the sockeye, *O. nerka*, also called the red or blueback salmon, weighing up to just 15 pounds (7 kg) and the pink or humpback salmon, *O. gorbuscha*, being a similar size. The coho or silver salmon, *O. kisutch*, grows to about 3¼ feet (1 m) and weighs up to 33 pounds (15 kg) as does the chum, *O. keta*, also known as the keta or dog salmon.

Color changes for spawning

Pacific salmon mainly return to spawn in their natal river, the same river in which they hatched. When they do so, breeding males change color. In the case of the pink and the chum salmon, the silvery sides become pale red with green-brown blotches, while the head and back often become darker. In the coho salmon, males become a darker blue green with a red stripe on the sides. Breeding males also grow long, hooked snouts. In most species the returning salmon are 4 or 5 years old. The pink salmon matures the earliest, at 2 years, the coho salmon at 2–4 years. However, some of the sockeye and chinook salmon may be as much as 8 years old before they return to fresh water to spawn.

Do not feed in fresh water

The salmon return in early summer, even in late spring or in the fall, depending on species and location. They head for the coast, away from their feeding grounds out in the Pacific, and stop feeding as they near fresh water and their digestive organs deteriorate. On reaching the mouth of a river, Pacific salmon head upstream, except the chum, which usually spawns near tidal waters. The coho salmon moves only a short distance upstream. The chinook, on the other hand, has been known to travel as much as 2,250 miles (3,620 km) up rivers. One exception to this is a subspecies of the sockeye that is nonmigratory. In contrast with the Atlantic salmon, Pacific

CHUM SALMON

CLASS **Osteichthyes**

ORDER **Salmoniformes**

FAMILY **Salmonidae**

GENUS AND SPECIES ***Oncorhynchus keta***

ALTERNATIVE NAME
Keta salmon; dog salmon

WEIGHT
Up to 33 lb. (15 kg)

LENGTH
Up to 3¼ ft. (1 m)

DISTINCTIVE FEATURES
In sea: steel blue or blue green above; silvery sides; white below. Breeding male (in fresh water): dark head and back; pale red sides with green-brown blotches.

DIET
Marine crustaceans such as copepods, tunicates and euphausiids; also mollusks, squid and small fish. Adult ceases feeding in fresh water.

BREEDING
Age at first breeding: usually 3–5 years; breeding season: depends on location; number of eggs: 700 to 7,000; hatching period: about 8 weeks; breeding interval: all adults die after spawning

LIFE SPAN
Up to 6 years

HABITAT
Adult: oceans for most of life; coastal streams to spawn. Migrating fry (juvenile): estuaries, close to shore, before entering sea.

DISTRIBUTION
North Pacific: Korea, Japan, Okhotsk Sea, eastern Russia and Bering Sea; Arctic Alaska south to San Diego, California

STATUS
Common

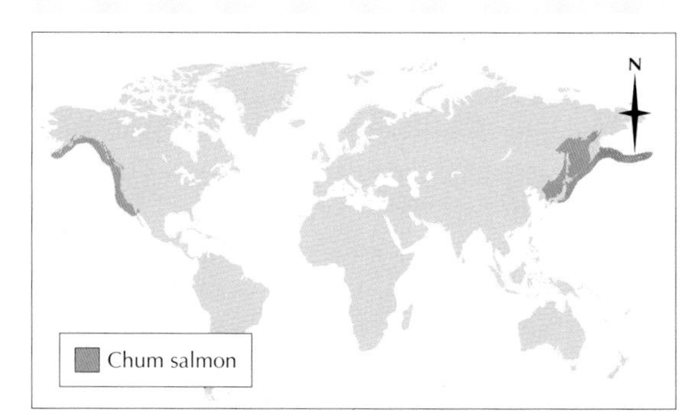

Sockeye salmon on migration along the Pacific coast of North America. The males of some Pacific salmon species become a brilliant red color as they migrate to fresh water for spawning.

salmon almost never survive the spawning run, although they may live for a few weeks after spawning. In only the pink salmon species do some of the adults survive and return to the sea.

Adults die after spawning

Because they are no longer feeding, the salmon are usually very thin as they near the spawning grounds. The males often look worse because they become aggressive, driving off other males and fighting with each other during this time.

The females, meanwhile, look for a place in the sandy or gravelly shallows where the water is clear and has plenty of oxygen. Then they start digging redds (troughs) in the riverbeds with their tails, each one lying on her side and flapping her tail. When the redd is deep enough, she drops into it to spawn, her mate swimming over to her to shed his milt (sperm-containing fluid), fertilizing the eggs. Each female lays several batches of eggs, 1,200 to 1,800 in the pink or humpback salmon, 700 to 7,000 in the chum salmon. Batches are laid in different redds, and by the end of spawning the female is completely exhausted. With her tail fins worn to stubs, her skin blackening with blotches of gray fungus, she dies. Males will often seek out other females, but eventually share the same fate. The carcasses of both sexes drift downstream or are stranded at the edge of the river or stream.

Sockeye salmon return to the Adams River, a branch of the Fraser River, British Columbia, to spawn in October. It is not known exactly how Pacific salmon navigate their way back to the rivers where they were hatched.

Down to the sea as infants

The female Pacific salmon cover each batch of eggs with sand and gravel. Thus protected, the orange-pink eggs hatch 8 weeks later. The alevins, or young salmon, remain under the gravel feeding on their yolk sacs for some weeks before wriggling to the surface as fry (juvenile fish). They feed on water fleas and other small animals and in the following spring are carried downstream by the current. The pink salmon and chum go to the sea as fry, but the sockeye may go as fry or as 1–3 year old fish, and the chinook and coho go when 1–2 years old.

Finding their way home

There has always been a great interest in how salmon find their way back to the streams where they were hatched. There is now evidence to show that the thyroid gland plays a part in the salmon's changing preference for water of varying salinity. When a coho was injected with a certain hormone, it sought seawater. When the injections were stopped, it sought fresh water. The opposite effect was found in the pink salmon. Other glands are probably involved, as well as the length of day and possibly the diet. The sense of smell may also play a part, as it does in finding food. In addition, temperatures influence the fish, certainly once they have entered fresh water. When temperatures are too low or too high, the fish make no effort to surmount obstacles. There is some evidence also that celestial navigation, using the sun by day and the stars by night, as in migrating birds, keeps the salmon on their runs along the coast to the mouths of the rivers in which they were spawned.

Controlled fishing

Many people living a long way from the Pacific are familiar with the Pacific salmon, but in canned form. The salmon fishery is commercially highly valuable. Sockeye salmon, for example, are taken in gill nets, reef nets and purse seines on their way to the Fraser River in British Columbia. Unrestricted fishing could kill the industry, so by an agreement between Canada and the United States, 20 percent of each subspecies of fish are allowed through to continue their journey to the spawning grounds. This is taken care of by a joint International Pacific Salmon Fisheries Commission, which also arranges for the catch to be divided equally between the two countries. There also is cooperation in providing concrete and steel fishways to assist the salmon up the rivers. The Pacific salmon fishery is close to an actual husbandry of a wild resource. Moreover, research is being carried out to produce strains of salmon that can tolerate less favorable rivers than they use at present and to transplant fry, which, when mature, will return to spawn in waters earmarked for cultivation.

PACK RAT

Although called rats, the pack rats, with their blunt muzzles, smallish eyes, fairly large ears and hairy tails, actually belong to the vole family. Their other names are trade rat and wood rat.

Pack rats are usually about 1½ feet (45 cm) long, of which about half is tail, and they can weigh up to 1 pound (450 g). They are dark brown to buff gray in color with white, gray or buff underparts. In some species the hairs are sparse on the tail, but in others the tail is quite thick and bushy.

The 20 species of pack rats are found from British Columbia, southern Canada, southward to Mexico. They are found throughout most of the United States. A typical species is the eastern or Florida pack rat, *Neotoma floridana.* The bushy-tailed wood rat, *N. cinerea,* is about 2 feet (60 cm) long, including its 8-inch (20-cm) tail. It ranges from British Columbia to California and eastward to the Dakotas. Found west of the Rockies, the dusky-footed wood rat, *N. fuscipes,* is the species that has been most studied.

A bushy-tailed wood rat, one of about 20 pack rat species, hibernating. This species is known for building large houses of sticks and leaves, often decorated with bright metal objects.

Thieves in the night

These nocturnal rodents seem to have an insatiable habit of picking things up and hoarding them. If on the way to its nest a pack rat sees something more attractive, it drops the object it is already carrying and takes the new one. Pack rats collect and hoard all sorts of objects, especially bright or colored objects or those made of metal. Because whenever they take something they tend to drop something else, it is almost as if an exchange has been made by the animal.

House builders

Some of the pack rat species build houses of sticks with many entrances and compartments. Some "rooms" are used for storing food, others for sleeping and one or more for garbage. The pack rats living in desert areas in the southwestern United States build their nests of pieces of cactus. They collect fragments of all kinds, including dead or living cactus, and pile them against a living cactus, to a height of 2 feet (60 cm) and several yards across. Inside the heap is a nest of dried grass and other soft materials. A maze of passages, protected by the hard, sharp cactus spines, leads up to it.

Hardly ever drink

Pack rats are found in all major habitats in the United States, including woodlands, swamps and rocky ground, but they seem to prefer scrub, desert and forest areas. When near lakes or streams, although they do not enter water, these rodents may build their nests over the water on or in fallen or leaning trees. In mangrove swamps they build in the trees.

Pack rats are mainly vegetarian, eating seeds, nuts, berries, leaves and roots, as well as a few small invertebrates. They drink little or not at all, getting their water from juicy plants. In deserts they get water almost entirely from the fleshy stems of cacti. Sometimes they live near farmhouses and can become a pest, feeding on agricultural crops.

Sometimes monogamous

Some pack rat species are said to be monogamous, the male remaining in the nest with the female even when she has young. The male courts the female by drumming with his feet, this behavior also being used as an alarm signal. In the northern parts of their range, pack rats mate

A desert wood rat, N. lepida. *Despite the name, which possibly comes from their habit of collecting and hoarding objects, pack rats actually belong to the vole family.*

in December and January and females give birth to only one litter each year. In the southern parts, on the other hand, breeding takes place year round, and there are two or three litters a year. Gestation lasts 30–40 days and there are one to eight young in a litter. The naked babies weigh ½ ounce (14 g) at birth. They begin to grow fur at 4 days and are fully furred in 2 weeks. The young are weaned after 3 weeks. At this point the mother leaves her family to occupy the nest while she goes off to find or build a new one.

Pack rats have a habit in common with a number of other rodents: that of licking saliva. The young lick the mother's saliva, possibly a way of transferring antibodies from mother to infant. Also, when two adult rats meet, they stop a short distance from each other, sniff at each other's noses and then come close to lick each other's lips, mouth, face and head. It is thought that this behavior may have some social purpose.

The dusky-footed wood rat

The species most extensively studied, the dusky-footed wood rat, is not monogamous as some other species are. The male mates with the female in the nearest nest, but tends to go from one nest to another. He is often driven from the nest when the female becomes pregnant. The nest of this species is made of sticks and has walls and a roof, is usually on two stories, with passages and chambers. To an extent the nests are owned communally, in that the rats often swap houses.

While most species of pack rats are still common, others are now at low risk. Along with six other species, subspecies of both the Florida wood rat and the dusky-footed wood rat are now considered endangered.

PACK RATS

CLASS **Mammalia**

ORDER **Rodentia**

FAMILY **Muridae**

GENUS ***Neotama***

SPECIES **20, including bushy-tailed wood rat, *Neotoma cinerea*; eastern pack rat, *N. floridana*; dusky-footed wood rat, *N. fuscipes*; and desert wood rat, *N. lepida***

ALTERNATIVE NAMES
Trade rat; wood rat; Florida pack rat (*N. floridana* only)

LENGTH
Head and body: 6–9 in. (15–23 cm); tail: 3–9½ in. (7.5–24 cm)

DISTINCTIVE FEATURES
Soft fur, although coarser in some species; brown or gray upperparts; white, cream or gray underparts; long tail may be bushy or almost naked, depending on species

DIET
Seeds, nuts, berries, leaves, roots and stems; occasionally small invertebrates

BREEDING
Age at first breeding: usually 7–8 months; breeding season: all year (southern regions), December–January (north); number of young: 1 to 8; gestation period: 30–40 days; breeding interval: 2 or 3 litters per year

LIFE SPAN
Average 3 years; up to 7 years in captivity

HABITAT
Mainly scrub, desert and forest areas

DISTRIBUTION
Southern Canada south to Mexico

STATUS
Most species common; others at low risk or endangered

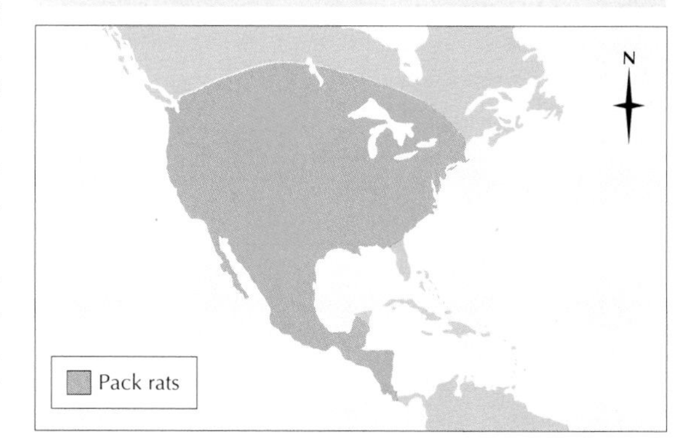

PADDLEFISH

THE PADDLEFISH, A LARGE freshwater bony fish, evolved nearly 65 million years ago, and it is supposed that some members of the family could have been present on Earth 300 million years ago, long before the dinosaurs roamed the planet.

The paddlefish has a long body, and looks rather like a shark. Its skin is naked except for a few scattered vestigial (imperfectly developed) scales and patches of scales on the tail fin. It has a fairly large head drawn out at the front into a long, flattened snout shaped like a paddle. This snout is between one-third and one-half the total length of the fish. At the base of the snout are the small eyes, and beneath them is a very wide mouth. The gill covers are large and triangular with the apex to the rear, and are drawn out into a point. The tail fin is only slightly forked.

There are just two species of paddlefish. One, the Mississippi paddlefish, *Polyodon spathula,* is also known as the spoonbill sturgeon. It lives in the Mississippi River system of North America, including the Missouri River into Montana, the Ohio River and their major tributaries. The other species is the Chinese swordfish, *Psephurus gladius.* It is found in the Yangtze River and its tributaries in China. The Mississippi paddlefish measures up to 7 feet (2.2 m) in length and weighs up to 200 pounds (91 kg). The Chinese species, sometimes called the swordbill sturgeon, is considerably larger, reported to reach some 10 feet (3 m) or more in length. It might weigh up to 660 pounds (300 kg).

Food detector and stabilizer

Some people say the paddlefish uses its elongated snout to probe in the mud for food, while others say it uses it to stir up the mud. This seems unlikely, however, for such a sensitive and easily damaged organ. More likely, as the paddlefish swims slowly along, it swings its highly sensitive paddle from side to side to detect its food. When it opens its large mouth, the back of the head seems almost to fall away from the rest of the body as the gill covers sag, revealing the capacious gill chambers. This sudden opening of the mouth and gill cavities probably produces a suction that draws in small plankton. As the fish swims forward, the plankton is strained from the water by the long gill rakers on the inner sides of the gills. In addition to plankton, paddlefish sometimes eat other fish, and shad, for example, have been found in their stomachs. More recently it has been discovered that the long, paddlelike snout also functions as a stabilizer in flowing water and reduces drag when the fish's mouth is open for feeding.

Rapidly growing young

Breeding takes place during spring and summer. If there are no dams to block their upstream progress, paddlefish might travel up to 200 miles (320 km) over the breeding period. After dark, a single female is courted by several males. A large female can produce up to 3 million eggs, laid in shallow water on top of flooded gravel bars. The eggs are ⅛ inch (3 mm) in diameter. and sink

A Mississippi paddlefish in the southeastern United States. It uses its long, elongated snout to detect food and as a stabilizer in flowing water. It is also able to sense electrical fields.

This preserved Mississippi paddlefish clearly shows the long, flattened snout, largely scaleless body and wide mouth of this unusual species. The paddlefish evolved nearly 65 million years ago.

down to the bottom where they attach to clean pebbles. The larvae hatch within 7 days at ordinary summer temperatures. They are then ⅓ inch (8.5 mm) long, with a large head, no eyes, no paddle and no barbels. Each feeds on a large yolk sac until it is used up. Spawning often takes place during a flood, and the larvae might be seen swimming erratically up to the surface and down again at this time. When the flood waters begin to subside they are carried downstream.

The larvae begin to grow their eyes and their barbels within a few hours of being hatched, but the paddle does not begin to grow for another 2 or 3 weeks. At first the paddle is only a small bump on the snout, but once it appears it grows rapidly. The young fish also grow quickly, reaching 1 foot (30 cm) in their first year. After 15 years, Mississippi paddlefish average 5 feet (1.5 m) long and weigh 42 pounds (19 kg). Males will not reach sexual maturity until 8 years old, however, while females mature at 10 years.

Relicts of the past

Paddlefish and sturgeons are the only surviving members of an order dating back millions of years, one which was an offshoot from the common ancestors of sharks and bony fish. The fact that there are only two species of paddlefish, one in North America and the other in widely separated eastern Asia, suggests that they are a dying race. If so, their end is being hastened, both in North America and in China, but not by fishing since they are not particularly valued as a source of food. Rather it is the building of dams and increasing levels of river pollution that have further restricted their range. For this reason both species of paddlefish are now under threat, with the Mississippi paddlefish classified as being at low risk, while the Chinese swordfish is critically endangered.

MISSISSIPPI PADDLEFISH

CLASS **Osteichthyes**

SUBCLASS **Actinopterygii**

ORDER **Acipenseriformes**

FAMILY **Polyodontidae**

GENUS AND SPECIES ***Polyodon spathula***

ALTERNATIVE NAME
Spoonbill sturgeon

WEIGHT
Up to 200 lb. (91 kg)

LENGTH
Up to 7⅕ ft. (2.2 m)

DISTINCTIVE FEATURES
Sharklike appearance; elongated snout, flattened from top to bottom; small eyes; large mouth; largely scaleless body; large, triangular gill covers with apex to rear; uniform lead gray in color

DIET
Mainly plankton; some other fish

BREEDING
Age at first breeding: 10 years (female), 8 years (male); breeding season: spring and summer, often during floods; number of eggs: up to 3 million; hatching period: about 7 days; breeding interval: 1 year

LIFE SPAN
Up to 30 years

HABITAT
Slow-flowing or still water in large rivers, usually at depths greater than 4¼ ft. (1.3 m); often found behind islands and sandbars

DISTRIBUTION
North America: Mississippi River system, including Missouri River into Montana, Ohio River and their major tributaries

STATUS
At low risk; uncommon in many areas

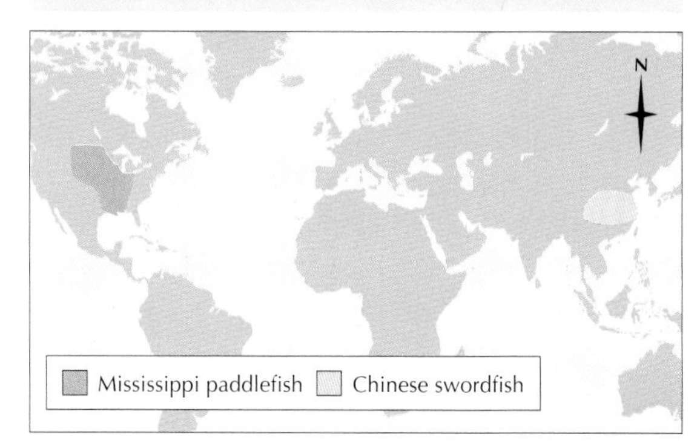

PADEMELON

THE NAME PADEMELON refers to a particular type of small wallaby. This word is said to come from the aboriginal name paddymalla, used to refer to small relatives of the kangaroos living in scrub. Confusingly, pademelons are sometimes called red-bellied wallabies.

Pademelons, of which there are six species, have a combined head and body length of up to 2½ feet (75 cm), with the tail growing up to 1½ feet (45 cm) long. The hind foot is not more than 6 inches (15 cm) long. The thick, rounded tail is less tapering than in other wallabies and is only sparsely coated with hair. The body fur is soft and thick. Pademelons are distinguished from other wallabies by the incisor teeth, their third upper incisor being broad with a notch at the rear edge.

The red-necked pademelon, *Thylogale thetis*, of the coastal strip of southern Queensland and New South Wales, is grizzled gray, reddish on the neck and shoulders and has a light stripe on the hip. The red-legged pademelon, *T. stigmatica*, of Queensland, New South Wales and New Guinea, is mainly russet with more reddish hind legs and a yellowish hip stripe. The red-bellied pademelon, *T. billardierii*, of Tasmania, is grayish brown with a rufous or orange front and a yellow hip stripe. The dusky pademelon, *T. brunii*, of New Guinea, is grayish brown to chocolate brown with a dark cheek stripe and a white patch above this and a yellowish hip stripe. The remaining two pademelon species, the New Guinea pademelon, *T. browni*, and Calaby's pademelon, *T. calabyi*, are also restricted to New Guinea.

Bolt-holes in scrub

Pademelons make tunnel-like runways in long grass, ferns and bushes. They live in thick scrub and dense forest undergrowth, especially in rain forest, forest dominated by eucalyptus, and swamp areas. The red-bellied pademelon of Tasmania lives in the gullies and scrub, where it makes well-worn paths. Pademelons sleep hidden in the undergrowth during the day, coming out at dusk to feed, mainly on grasses but also to some extent on young shoots, herbs and seedlings.

During the 20th century there was some scientific debate over the question of whether wallabies and kangaroos chew the cud as cattle do. This issue was investigated toward the end of the century by B. C. Mollison of the Commonwealth Scientific and Industrial Research Organisation in Australia. He found that the pademelon sometimes brings up food while resting, perhaps 1–2 hours after it has finished feeding. The regurgitation is preceded by vigorous heaving movements of the chest and abdomen and the food brought up is retained in the mouth and chewed thoroughly before it is swallowed again. Some of it is spilled from the sides of the mouth onto the ground and it is traces of this kind that indicate such rumination is a fairly constant habit. Mollison fed one of the wallabies with bread, biscuit, carrot and apple, but none of these were regurgitated, although grass and other herbage that the pademelon had eaten previously was brought up to be chewed. It seems therefore that wallabies and pademelons do not chew the cud in the manner of cattle, but rather give a second, more thorough chewing to fibrous foods such as grass in order to ensure that they are as completely digested as possible.

Decline of the species

At the time of the settlement of Australia by Europeans, pademelons were widespread. The aboriginal peoples native to Australia killed the animals for food but their actions had little effect

Although abundant in Tasmania, the red-bellied pademelon has not been seen on mainland Australia since the 1930s.

Pademelons shelter in dense vegetation by day, emerging at dusk to feed on soft grasses, herbs, tree seedlings and shrubs.

on pademelon numbers. Soon after their arrival in Australia, the early white settlers realized that pademelon flesh was tender and well-flavored, tasting like hare according to some reports. In a short while, these small wallabies were being hunted both for the pot and for their skins, which were used for rugs and for trimmings. Hunters snared thousands of pademelons for their pelts and the animals were also menaced by bush fires. The pademelons' survival was further hampered by the fact that they are slow breeders, with one young, rarely two, at a birth, although the female has four teats in her pouch.

Predators and rivals

Over time the pademelons' habitat was increasingly reduced by land clearance and fencing. This let in the larger wallabies and the kangaroos, which outgrazed the pademelons. Within historic times the Tasmanian wolf and the Tasmanian devil were common on the mainland of Australia. There were large numbers of marsupial cats (covered elsewhere in this encyclopedia), and the wedge-tailed eagle, now rare except in remote areas, was one of Australia's most common birds in the early days of European settlement. All these species, and probably others as well, fed on the herbivorous marsupials, particularly on the smaller ones, such as the pademelons. Introduced red foxes may have been the pademelons' chief predator, although domestic cats and dogs that escaped and turned wild also attacked them. The spread of the introduced rabbit in Australia gave the pademelons a major competitor for food and living space. Although pademelons are thought to be still common over parts of their former range, they have been much reduced in numbers, particularly in more settled parts. The red-bellied pademelon is no longer found in Australia.

PADEMELONS

CLASS **Mammalia**

ORDER **Marsupialia**

FAMILY **Macropodidae**

GENUS ***Thylogale***

SPECIES **Red-necked pademelon, *T. thetis*; red-legged pademelon, *T. stigmatica*; red-bellied pademelon, *T. billardierii*; dusky pademelon, *T. brunii*; New Guinea pademelon, *T. browni*; Calaby's pademelon, *T. calabyi***

ALTERNATIVE NAME
Tasmanian pademelon, rufous wallaby (*T. billardierii*); alpine wallaby (*T. calabyi*)

WEIGHT
4–26½ lb. (1.8–12 kg)

LENGTH
Head and body: 11½–26½ in. (29–67 cm); tail: 9⅔–20 in. (24.5–51 cm)

DISTINCTIVE FEATURES
Stocky build; thick gray coat with patches of redder fur in places; lighter stripe on hips in some species; relatively short tail

DIET
Mainly grasses; also herbs and seedlings

BREEDING
Age at first breeding: 14–18 months; breeding season: varies according to species and location; number of young: usually 1; gestation period: 30 days

LIFE SPAN
Up to 9 years in captivity

HABITAT
Forest, savanna, thick scrub and grassland

DISTRIBUTION
Eastern Australia and New Guinea; Tasmania

STATUS
***T. calabyi*: endangered; *T. brunii*, *T. browni*: vulnerable; other species locally common**

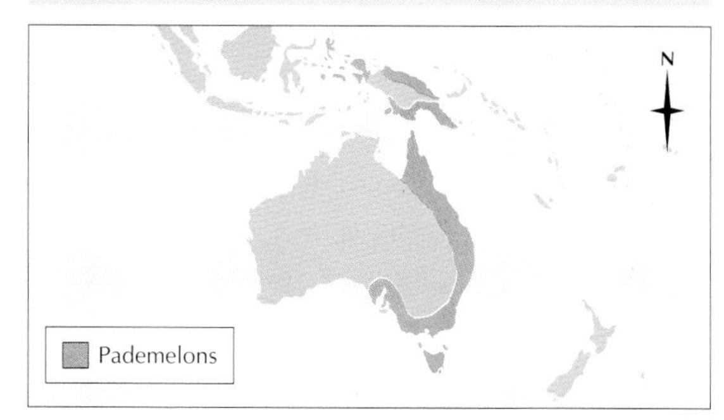

PAINTED SNIPE

Painted snipes bear only a superficial resemblance to true snipes. The former have bills that are hard, rigid and shorter than those of true snipes, and their flight is slow compared with the swift flight of true snipes. Painted snipes are closely related to plovers and sandpipers. The females take the more active part in courtship and are brightly colored, whereas the males are more drab.

There are two species of painted snipes. The Old World painted snipe, *Rostratula benghalensis*, which is 9–11 inches (23–28 cm) long, has a slightly curved bill and its eyes are well placed for stereoscopic vision. The ornately patterned females are dark green above, chestnut on the sides of the head and neck and white underneath. The males are brown barred with black above and white underneath. Both species have white spectacle-like rings around the eyes and yellowish spots on the wings and tail. The legs are green. The Old World painted snipe lives in much of sub-Saharan Africa, excluding the densest forests and the deserts, but including Madagascar. It is also found in the Nile Delta in Egypt, Asia south of the Himalayas, Japan, Indonesia, the Philippines and Australia.

The South American painted snipe, *Nycticryphes semicollaris*, is about 10 inches (25 cm) long. The bill is more curved at the tip than that of the Old World species and the toes are slightly webbed. The plumage of the sexes is similar although that of the female is brighter. It is almost black above with buff stripes and two oval white spots on the back, the underparts being pure white.

Painted snipes eat worms, snails, insects, seeds and vegetation such as rice paddy. In Japan, some birds eat seeds almost excusively.

Sleepy heads

Painted snipes can be found in flocks of up to 40, but are usually found singly or in pairs, living solitary lives in swamps and marshes or in grassland. They are nocturnal or crepuscular (active at twilight) and when disturbed they crouch, motionless, only taking flight as a last resort.

The flight of painted snipes is weak, and when they are flushed they merely flutter a short distance away from the source of the disturbance, with their legs dangling, before landing. In Argentina *N. semicollaris* is known as dormilón, meaning "sleepy head," a reference to this apparently lethargic behavior.

Although she is not usually noisy, the female painted snipe is able to utter a resonant boom, produced in a greatly elongated trachea, twice the length of the neck. By contrast, the male has a straight trachea and can only give a weak chirp.

In contrast with most other species of birds, the female painted snipe (*R. benghalensis*, above) is larger and more colorful than the male, and she initiates courtship.

Matriarchal society

As with other birds, such as phalaropes, in which the female's plumage is brighter than that of the male, the female painted snipe holds the territory and takes the initiative in courtship, leaving the male to bring up the family. The Old World painted snipe is polyandrous (the female has many mates). Each female may have several partners, with nests 13–33 feet (4–10 m) apart, but she may lay in nests up to 600 feet (183 m) apart, starting each new clutch at intervals of 12 days.

Nesting occurs during the rainy season, but in some places breeding may take place year-round, sometimes twice a year. The nest is built on wet ground. The Old World painted snipe makes an untidy mass of grasses and roots to keep the eggs above water, and sometimes the reeds and long grasses that surround the nest are woven together to make a canopy. The South American painted snipe lays its eggs on damp ground among herbage, often without a lining.

Painted snipes perform a broken-wing distraction display (above), pretending to be injured so as to attract the attention of a predator and draw it away from the nest.

The Old World painted snipe lays four eggs, sometimes two or three. They are cream or yellowish with black blotches and speckling, and the male takes full responsibility for their incubation. The American species lays only two eggs, which are white with black specks, and the female takes part in the incubation of her eggs, which lasts about 19 days. The chicks leave the nest shortly after hatching. If it is disturbed, the sitting bird does not leave the nest until the intruder is very near. Then it runs silently away or performs a broken-wing distraction display to distract the predator's attention from the nest.

Dual display

Female painted snipes employ a spectacular display for their courtship ritual and also for threatening intruders or rivals. They spread their wings and bring them forward on either side of the bill, also raising the tail vertically and fanning it. Throughout the display, the bird hisses. This performance reveals the bright spots and marking on the wings and tail to dramatic effect. The females also preen the males as part of courtship. Female painted snipes display to territorial rivals, by facing them and spreading their wings, and fight aggressively over partners.

If it is threatened by a predator, a painted snipe stops a few feet from the intruder, spreads its wings and tail with a rustle of feathers and poses, swaying backward and forward. Then it charges and jabs with its bill. There is no weight behind the jab but the display is frequently impressive enough to dissuade the intruder from searching for the snipe's brood.

OLD WORLD PAINTED SNIPE

CLASS **Aves**

ORDER **Charadriiformes**

FAMILY **Rostratulidae**

GENUS AND SPECIES ***Rostratula benghalensis***

WEIGHT
Male: 4¾–5 oz. (120–145 g); female: 4½–6¾ oz. (130–190 g)

LENGTH
Head to tail: 9–11 in. (23–28 cm); wingspan: 19¾–21¾ in. (50–55 cm)

DISTINCTIVE FEATURES
Long, slightly decurved bill; longish green legs; female up to 10 percent larger than male; white spectacle-like rings around eyes; white belly. Female: chestnut face and neck; dark green upperparts. Male: brown neck, face and upperparts.

DIET
Seeds, worms, snails and insects; also some vegetation

BREEDING
Age at first breeding: 1 year (male), 2 years (female); breeding season: egg-laying all year in Tropics, more restricted season in higher latitudes; number of eggs: usually 4; incubation period: 19 days; breeding interval: female mates with several males, with very short interval between different clutches

LIFE SPAN
Not known

HABITAT
Swamps with muddy patches, shrubby vegetation or reed beds; also rice paddies

DISTRIBUTION
Much of sub-Saharan Africa; Madagascar; Nile Delta, Egypt; southern Asia; Australia

STATUS
Locally common

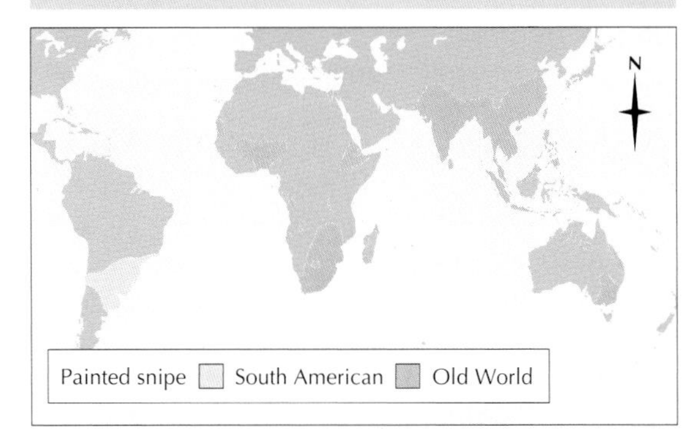

PAINTED TURTLE

THE PAINTED TURTLE IS the most widely distributed, and in many places the most common, small turtle in North America. Its smooth, flat carapace (shell) is usually 4–7 inches (10–18 cm) long, although the largest painted turtle ever recorded had a shell length of 9¾ inches (25 cm). The markings are particularly handsome, and the name of painted turtle is well deserved. The carapace is olive green to black, with yellow transverse bands and bright red markings around the edge. The red markings are less conspicuous in older turtles. The legs may be marked with red lines and the head with horizontal yellow lines. The plastron (the underside of the carapace) is bright yellow and looks as if it has been freshly scrubbed.

The sole species of painted turtle, *Chrysemys picta*, has four subspecies living in different regions of the United States, in southern Canada and in a very small area of northern Mexico. The eastern painted turtle, *C. p. picta*, is found on the Atlantic coastal plain and the southern painted turtle, *C. p. dorsalis*, in the Mississippi valley. Both have a plain yellow plastron. The Midland painted turtle, *C. p. marginata*, lives between the Allegheny Mountains and the Great Lakes. Its plastron has black markings confined to the center. The most widely distributed subspecies is the western painted turtle, *C. p. belli*, which has black markings around the edges of all the plates of the plastron. It is found in the plains of the northern United States and southern Canada and along the east of the Rockies and in parts of northern Mexico.

The distribution of these four distinct subspecies is believed to have resulted from the migrations of their ancestors during the last Ice Age, when different groups became separated in different geographical areas. When the turtles moved northward again after the Ice Age, some of these differences persisted. Closely related to the painted turtle are the small turtles of the genus *Pseudemys*, known as cooters and sliders. The habits of the two genera are similar.

Home-loving turtles

When food is plentiful, painted turtles do not move far, each having a home range that is about 100 yards (90 m) in diameter. This is about the distance from which an individual can successfully find its way home, probably by means of landmarks, if it has been displaced. When paint-

Painted turtles basking in the sun. They do this for several reasons, including the need to maintain their body temperature and to synthesize vitamin D_3, vital for processing calcium in their bodies.

Painted turtles are fairly small turtles and are sometimes called painted terrapins. The western subspecies tends to grow larger than the other three.

PAINTED TURTLE

CLASS **Reptilia**

ORDER **Testudines**

FAMILY **Emydidae**

GENUS AND SPECIES ***Chrysemys picta***

ALTERNATIVE NAMES
Painted terrapin; basking turtle; basking terrapin; eastern painted turtle (*C. p. picta*); western painted turtle (*C. p. belli*); southern painted turtle (*C. p. dorsalis*); Midland painted turtle (*C. p. marginata*)

LENGTH
Up to 9¾ in. (25 cm); most adults 4–7 in. (10–18 cm)

DISTINCTIVE FEATURES
Smooth, flat carapace (shell); coloration varies according to age and subspecies; olive green or black upper side to carapace, with yellow transverse bands and red markings at edge; plastron (underside of shell) often bright yellow; yellow markings on head; sometimes also red lines on legs

DIET
Leaves of water plants and invertebrates; also tadpoles, small fish and some carrion

BREEDING
Age at first breeding: at least 4 years; breeding season: late May–mid-July; number of eggs: 4 to 20; breeding interval: 1 year

LIFE SPAN
Not known

HABITAT
Slow-moving rivers and streams; shallow ponds and lakes

DISTRIBUTION
U.S. except south and southwest; southern Canada; small area of northern Mexico

STATUS
Very common

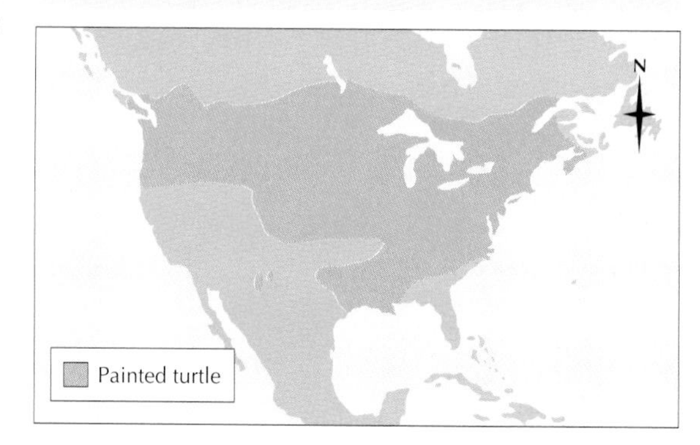

ed turtles set off in search of food, they move apparently at random, although they may be able to tell where water is from the brightness of the light that is reflected from it.

Painted turtles live in ponds, shallow parts of lakes and other quiet stretches of water. They may be found in the sluggish, well-vegetated parts of streams. Their distribution is determined by the abundance of water plants, their chief food. They are particularly fond of long, trailing plants that float on the surface. Aquatic insects, including beetles and dragonfly larvae, snails, tadpoles and fish are also eaten. If food runs low—for example, in a drought—painted turtles migrate to other stretches of water.

Besides emerging to go in search of fresh food supplies, painted turtles rarely come on land. Like all turtles, however, they lay their eggs on land. Painted turtles also climb onto banks or

logs to bask during the morning and afternoon; they avoid the midday sun. It has been suggested that one benefit of basking is that it dries the skin, causing parasitic leeches to drop off.

In the spring and early summer the male, distinguished by his longer tail and claws on his forefeet, seeks out and courts the female. He swims after her, overtakes her and turns to face her head-on. The female continues swimming, so the male has to swim backward, but as he goes he brushes the long claws of his forefeet against the female's cheeks. If a female is receptive, she sinks to the bottom of the pond and allows the male to mate with her.

A dash for safety

Eggs are laid from May to July. The females crawl onto land and dig their nests up to 100 yards (90 m) or more from the water's edge. They dig the nest with the hind feet and sometimes soften the soil with urine. The females carefully position each egg in the nest with their hind feet. A clutch averages 4 to 20 eggs and after the female turtle has finished laying she fills in the hole and stamps down the soil. The eggs hatch in the autumn, except in the most northerly part of the range, where they may not hatch until the following spring. The newly hatched turtles are about 1 inch (2.5 cm) long. They try to make their way to the water almost immediately, before they can be caught by their numerous predators. They are mature at 4 years, by which time they are 3½ inches (9 cm) long.

Seeking warmth

In the northern parts of their range, painted turtles hibernate, burrowing into mud or sheltering in a muskrat hole from November to March. When they emerge in the spring, their pond may still be covered with ice. It has been found that painted turtles sometimes move some distance from their "home" pond at these times and travel up the streams that flow in or out of it.

To discover what, if anything, guided the turtles, scientists decided to place some of them in Y-shaped tubes. The turtles could choose which arm of the Y to swim up, and it was found that they responded to the temperature of the water in the arms. Turtles do not become active until the temperature reaches 50° F (10° C). If the water in which they had been living was too cool, they swam up the warmer arm, but if accustomed to warm water, they followed the cool current. In the spring, then, the turtles sometimes leave their ponds if they are still too cold and swim up the slightly warmer streams. As their bodies warm up, the turtles become more active and begin feeding.

The various subspecies of painted turtles can be difficult to tell apart. The bright yellow spots on the head identify this individual as an eastern painted turtle.

PALMCHAT

The palmchat's name derives from its distribution, which matches that of the royal palm tree on the island of Hispaniola.

THE PALMCHAT IS NATIVE to the Caribbean island of Hispaniola, composed of the Dominican Republic and Haiti, and is also found on nearby Gonave and Saoma Islands. It is one of the most prominent birds on Hispaniola, partly due to the large communal "apartment houses" it builds. The palmchat is a noisy and gregarious bird. It is 8 inches (20 cm) long and is olive brown above, while its buff-yellow underparts are heavily streaked with brown. The rump is dark green, the tail is fairly long and the wings are rounded. The sexes are similar and juveniles differ from adults in having dark brown necks and light brown rumps. The palmchat is the sole representative of the family Dulidae. It is also the national bird of the Dominican Republic, where its flocks congregate in the royal palm savannas. It feeds mostly on fruits and palm berries.

Apartment houses

In the spring, from March to June, the palmchats build their nests or repair old ones. The nests are woven from twigs that are 1 to 2 feet (30–60 cm) long and up to 1 inch (2.5 cm) in cross section. The birds carry the twigs in the bill from the ground to the nest, as much as 50 feet (15 m) high in a palm tree. They weave the twigs loosely at first, then tighten them to form an untidy mass up to 7 feet (2 m) in diameter. The nest is securely anchored to the palm fronds, but when these eventually die it falls to the ground.

PALMCHAT

CLASS **Aves**

ORDER **Passeriformes**

FAMILY **Dulidae**

GENUS AND SPECIES ***Dulus dominicus***

WEIGHT
About 1½ oz. (40 g)

LENGTH
Head to tail: 8 in. (20 cm)

DISTINCTIVE FEATURES
Olive-brown upperparts; heavily streaked buffish underparts; stout bill; reddish iris

DIET
Fruits and berries, particularly of palm trees

BREEDING
Age at first breeding: 1 year; breeding season: eggs laid March–June; number of eggs: 2 to 4; breeding interval: 1 year

LIFE SPAN
Not known

HABITAT
Royal palm savanna and other open areas with scattered trees

DISTRIBUTION
Hispaniola

STATUS
Common

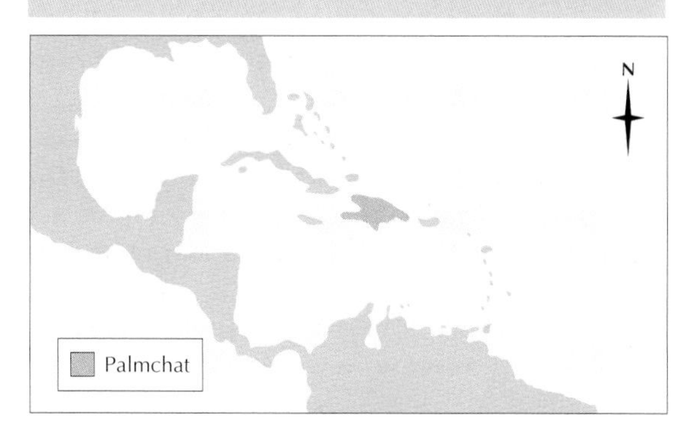

Several pairs of palmchats live in one nest, which they build together. Each nest is divided into separate compartments, in which each pair raises its family. The nest is made up of units each consisting of a chamber, 4–5 inches (10–12.5 cm) across, with tightly woven walls and a tunnel leading to the exterior. Smaller tunnels lead inside the nest so the birds can move about under cover. The nesting chambers are lined with shredded bark and grass and the nests are also used as roosts outside the breeding season.

PALM CIVET

THERE ARE 11 SPECIES of palm civets in southern Asia and one in Africa. The 17 species of civets are discussed in a separate article.

Palm civets are long-bodied and short-legged, with a sharp muzzle and a tail as long as or slightly longer than the head and body. The best-known species is the common palm civet, *Paradoxurus hermaphroditus,* also called the musang, or toddy cat. It ranges from Sri Lanka and India through southern China to Southeast Asia. Three other palm civets of the same genus live in southern India and Sri Lanka. Just over 4 feet (1.2 m) in total length, all three species are gray to brown with dark back stripes, spots on the flanks and white patches on the head.

The Javan small-toothed palm civet, *Arctogalidia trivergata,* which ranges from Assam south to Borneo, is much the same length but with a longer tail. It is grayish to orange or tawny, with dark stripes along the back and a white stripe on the muzzle. The species gets its name from its very small back teeth. The similar but more slender Sulawesi palm civet, *Macrogalidia musschenbroeki,* is rare and confined to that island. It is 5 feet (1.5 m) long, has more gray on the face, throat and belly than the small-toothed palm civet and has a ringed tail. The masked palm civet, *Paguma larvata,* is 4½ feet in total length and lives in Kashmir, from southern China eastward to Hainan and Taiwan and on the Andaman Islands. It has also been introduced to Japan.

The African palm civet, *Nandinia binotata,* is about 4 feet (1.2 m) overall and ranges from Senegal east to the Sudan and south to Zimbabwe and Angola. It is gray to brown tinged with buff or chestnut, with lighter underparts. The African palm civet sometimes has two white spots on the shoulders, from which it gets its alternative name of two-spotted civet.

Banded palm civets

The remaining three species of palm civets all inhabit Southeast Asian rain forests and are grouped together as the banded palm civets. The Malay banded palm civet, *Hemigalus derbyanus,* is brown and banded with dark rings, while Owston's palm civet, *Chrotogale owstoni,* has spots as well as rings. The Borneo banded palm civet, *Diplogale hosei,* is blackish brown all over.

Whereas banded palm civets are terrestrial or spend only a little time in trees, other palm civets are mainly arboreal (tree-living). In climbing, the latter palm civets are helped by their sharp, curved claws and, in most species, by the naked soles of their feet. All palm civets are active at night, the pupil of the eye being vertical, as it is in many other nocturnal animals.

Little is known about palm civet calls, since most of the species are fairly silent animals. Some use a kind of mewing, with light snarls coupled with hissing and spitting when alarmed. During the day palm civets rest in holes in trees, among vines or in other dense foliage. Common palm civets often live in or near houses, in gardens, in dry drains or in thatch, even venturing into the suburbs of large towns.

Carnivores that eat fruits

Although similar to the true civets of the subfamily Viverrinae in appearance and habits, the palm civets differ from them in several anatomical details, including the teeth. Palm civets' teeth show these animals to be less carnivorous than the true civets. Generally, the teeth are weaker than those of true civets, and the carnassials, the

A Malay banded palm civet peeks through the foliage of its Sumatran forest home. Although palm civets are true carnivores, they eat mainly fruits and seeds.

A common palm civet scuttles down the bough of a tree. Unlike most palm civets, this species is often seen near human habitation.

teeth used for slicing flesh, are less well developed in the palm civets than in more typical carnivores. Most palm civets eat mainly plant food, especially fruits, but also seeds, and even roots in the case of the masked palm civet. To this they add such food as small birds, eggs, insects and their grubs, tree sap and palm juices.

The masked palm civet also catches fish and has the reputation of being a good ratter. The African palm civet goes for even bigger prey, such as nocturnal pottos, *Perodicticus potto*. The three species of banded palm civets tend to eat more animal food than do the other palm civets, the Malay banded palm civet being particularly predatory and Owston's palm civet tending to specialize in hunting invertebrates.

The *Paradoxurus* species' alternative name of toddy cats comes from their liking of fermenting palm sap, or toddy. The local peoples tap the palm trees, hanging cups on the trunks to catch the sap for wine-making. Before this is collected the palm civets often take a share.

There are two to five young in a palm civet litter, born in a hollow tree or among rocks. Females usually have a litter every 1 or 2 years, and gestation lasts 60–64 days. The young grow to full size in about 3 months. Palm civets, especially the Indian species, are often kept in zoos, where they have lived to 15 years of age.

Skunklike tactics

Some of the palm civets use their anal glands as a means of communication and defense. The four species of the genus *Paradoxurus* especially, and to a lesser extent the masked palm civet, give out an obnoxious odor from these glands, which is more in keeping with the skunks.

PALM CIVETS

CLASS **Mammalia**

ORDER **Carnivora**

FAMILY **Viverridae**

GENUS **8 genera**

SPECIES **11 species, including common palm civet, *Paradoxurus hermaphroditus*; African palm civet, *Nandinia binotata*; and Malay banded palm civet, *Hemigalus derbyanus***

ALTERNATIVE NAMES
***P. hermaphroditus*: musang; toddy cat; *N. binotata*: two-spotted civet**

LENGTH
***N. binotata*. Head and body: 17⅓–23 in. (44–58 cm); tail: 18–24½ in. (46–62 cm).**

DISTINCTIVE FEATURES
Long, lithe body; pointed muzzle; long, well-furred tail. Most species: gray to brown, with variety of spots and stripes. Banded palm civets: dark rings around body.

DIET
Mainly fruits and seeds; some insects, tree sap, eggs and birds; also palm juices, fish and small mammals (certain species only)

BREEDING
Age at first breeding: 3 years; number of young: 2 to 5; gestation period: 60–64 days; breeding interval: usually 1–2 years

LIFE SPAN
Up to 15 years in captivity

HABITAT
Forests and plantations

DISTRIBUTION
***N. binotata*: sub-Saharan Africa; other species: India east to southern China, south to Malaysia and Indonesia.**

STATUS
Vulnerable: at least 4 species; most other species locally common

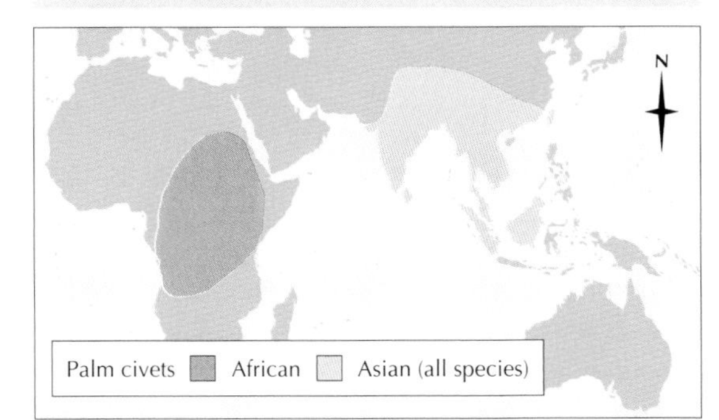

PALM DOVE

THE PALM DOVE IS A relative of the collared dove, *Streptopelia decaocto*. It is well known in many parts of Africa and Asia and has several alternative names, including laughing dove. The palm dove is much smaller than the domestic pigeon, with comparatively short wings and a longish tail. The head, neck and breast are a dark mauvish pink merging into the white of the belly, and the bill is dark. The back is reddish brown, the rump slate blue and the tail brown, edged with gray and white. The wings are largely slate blue. Around the neck there is a bib of colorful feathers, each forked and black at the base with golden tips. The female palm dove is paler than the male.

The palm dove is found in much of Africa, including the gallery forests and savanna of East Africa, though not in the Sahara region or Madagascar. In Asia it ranges from Turkey and the Middle East eastward to Bangladesh and western China. Palm doves also live on Malta and have been introduced into parts of Australia. Those found in Turkey and Central Asia were probably introduced by humans.

Palm doves rely heavily on water and remain close to any area that has a water source.

Laughing calls

Palm doves favor open country such as dry scrub and bush where permanent water is near, but they are also found in forest clearings. They have come to associate with human habitations in many places, a fact that has contributed toward their alternative common names of town dove, village dove and garden dove. They are very tame and feed largely on spilled seeds such as millet, wheat and maize. Their natural food is the seeds of several wild plants, but they will also take snails and insects, particularly termites.

Although they live in pairs, several dozen palm doves can often be seen feeding together in streets, on farms or on the rocks where cassava is pounded. When they are feeding, groups of laughing doves frequently argue among themselves, hooting and moaning quietly and occasionally adding force to their bickering with pecks. The flight of palm doves is usually slow but they can fly very quickly if chased by a predator, such as a hawk. Their main call is a hollow *ha-ha-hoo-hoo-hoo-hoo-hoo* that sounds like laughter. This characteristic call has given rise to the bird's alternative name of laughing dove.

Billing and cooing

Like other doves, the male palm dove has a spectacular display flight: it flies up, claps its wings and then glides in a circle, gradually descending with its wings and tail spread. The display flight is used to advertise the male's presence to females and other males. When courting, the male bows rapidly to the female, with his head horizontal, which enables him to show off the colorful bib on his neck. At the same time he coos softly. If the male palm dove's bowing is acceptable to the female, the pair engages in billing, which leads to mating.

In some species of doves billing simply involves bill touching, but the gesture is derived from courtship feeding. Prior to mating, the male offers his open bill, the female inserts hers and food is transferred. In the palm dove no food is passed between the birds but the male makes a gesture as if he is regurgitating food. Following copulation, the male stands erect with his neck feathers puffed out and his head up. At the same time the female parades with the feathers on her rump and neck erect.

The palm dove spends much of its time on the ground, moving with short, shuffling steps. It feeds mainly on small seeds and grains, and some insects.

Light but strong nests

In the Tropics palm doves nest at any time of the year, but most of the eggs are laid during the rainy season. In North Africa palm doves breed between February and October. The birds build their nests in the forks of trees or at the bases of palm fronds and sometimes on ledges of buildings or in thatch. The male collects the material while the female builds. As is usual with pigeons and doves, the nest appears to be very flimsy and hardly able to support the weight of the two white eggs and the sitting parent. Both male and female palm doves share in the incubation of the eggs. At times the eggs show through the bottom of the nest, a fact that might suggest little care is taken with nest construction. However, in his book *Birds In My Indian Garden*, published in 1960, the ornithologist Malcolm MacDonald refuted the notion that the pigeon family assemble their nests carelessly. A palm dove nest in his garden lasted for at least 7 months, even surviving the monsoons. When the author took several abandoned palm dove nests apart he discovered that they were made of 80 to 140 twigs and rootlets that were woven into a firm pad.

Young palm doves are known as squabs. Their parents feed them "pigeon's milk," which is a mixture of partially digested food and a secretion from the walls of the parent bird's crop. Squabs have brownish heads and are duller in color than the parents, and do not have the characteristic chest markings of adult palm doves.

PALM DOVE

CLASS **Aves**

ORDER **Columbiformes**

FAMILY **Columbidae**

GENUS AND SPECIES ***Streptopelia senegalensis***

ALTERNATIVE NAME
Laughing dove

WEIGHT
2¾–3¾ oz. (77–105 g)

LENGTH
Head to tail: 10–10⅔ in. (25–27 cm); wingspan: 15¾–17¾ in. (40–45 cm)

DISTINCTIVE FEATURES
Red-brown and blue-gray upperparts; dark mauve-pink head, neck and breast, merging into white belly; black spots on upper chest; white tip to tail; wings appear fairly short in flight; relatively long tail

DIET
Mainly seeds; occasionally insects, particularly termites, and small snails

BREEDING
Age at first breeding: 1 year; breeding season: eggs laid all year in Tropics, February–October in subtropics; number of eggs: 2; incubation period: 12–14 days; fledging period: 14–17 days; breeding interval: usually 2 broods per year

LIFE SPAN
Up to 6 years

HABITAT
Cultivated areas, towns and villages, scrub, bush, woodland edges and oases

DISTRIBUTION
Much of Africa; parts of Middle East eastward to Bangladesh and western China; introduced to state of Western Australia

STATUS
Locally abundant

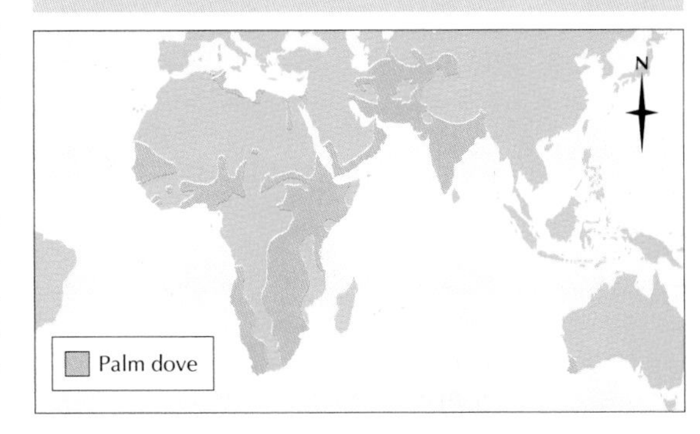

PALM SQUIRREL

THERE ARE 10 SPECIES OF squirrels with this name, belonging to four separate genera. The best-known are the palm squirrels of southern Asia, genus *Funambulus*, which are found across Sri Lanka, India, Pakistan and the Andaman Islands. The five species of Asian palm squirrels are often called striped palm squirrels because they are striped much like chipmunks. They have a dense, soft fur that is light grayish brown to almost black most of the year but shows reddish on the head from December to May. They have three light stripes on the back, occasionally with a further short stripe on each flank. Asian palm squirrels are around 7 inches (18 cm) long in head and body, with about the same length of bushy tail.

The African palm squirrels include two species, the western palm squirrel, *Epixerus ebi*, and the Biafran Bight palm squirrel, *E. wilsoni*. They are up to 2 feet (60 cm) long, of which about half is a bushy tail. The Biafran Bight palm squirrel lives in Gabon, Cameroon and Democratic Republic of Congo (Zaire). The coat color is a mixture of red and black and is yellowish on the underparts, where the hair is often scanty and the skin at times almost naked. The western palm squirrel, found in Ghana and Sierra Leone, is reddish with buff patches and some yellow on the underparts. Again, the hair on the underparts is sparse, the skin often naked.

A third group are known as the oil-palm or African giant squirrels. There are two species. One, the slender-tailed squirrel, *Protoxerus aubinnii*, is found in West Africa from Liberia to Ghana, and the other, the African giant squirrel, *P. stangeri*, lives farther south, from Kenya to Angola. The first has grizzled fur and a slender black tail, whereas the second is tawny olive to nearly black, shading to white or buff on the underparts, with white to grayish cheeks. Its bushy tail is sometimes banded black and white. The length of head and body is up to 15¾ inches (40 cm); the tail is 14 inches (36 cm) long. There is also a single species of the genus *Menetes*, Berdmore's palm squirrel, *M. berdmorei*.

Varied habitats

The different species of Asian palm squirrels live in very varied habitats, from open palm forest and scrub at low altitudes to dense jungle with tall trees. Some species are also found at high altitudes. Their habits are similar to those of tree squirrels of the Northern Hemisphere, and, like them, some palm squirrels live near human settlements, sometimes even nesting in roof spaces. African palm squirrels usually live in cavities in trees. The African giant squirrel is also known as the booming squirrel because of the booming sounds it makes when alarmed. Palm squirrels are generally common, but some species are rarely seen because they hardly ever come down to the ground.

Act as pollinators

Asian palm squirrels feed by day, in the trees or on the ground. They mainly eat nectar, pollen, fruits, seeds and seed pods, as well as insects and their grubs. Some species specialize on palm pods. When taking nectar from the silky oak, which is not a true oak, the squirrels become dusted with pollen and so may act as pollinators. African palm squirrels feed in trees and on the ground, especially on the nuts of the oil palm. They also eat carrion and gnaw on bones.

Fight over females

At breeding time a female is chased by several males, which fight among themselves while she waits on a nearby tree making peeping calls at her fighting suitors. The successful male then

A three-striped palm squirrel,* Funambulus tristriatus, *in a fruit-bearing tree, Karnataka, India. Some species of palm squirrels spend most of their time in trees.

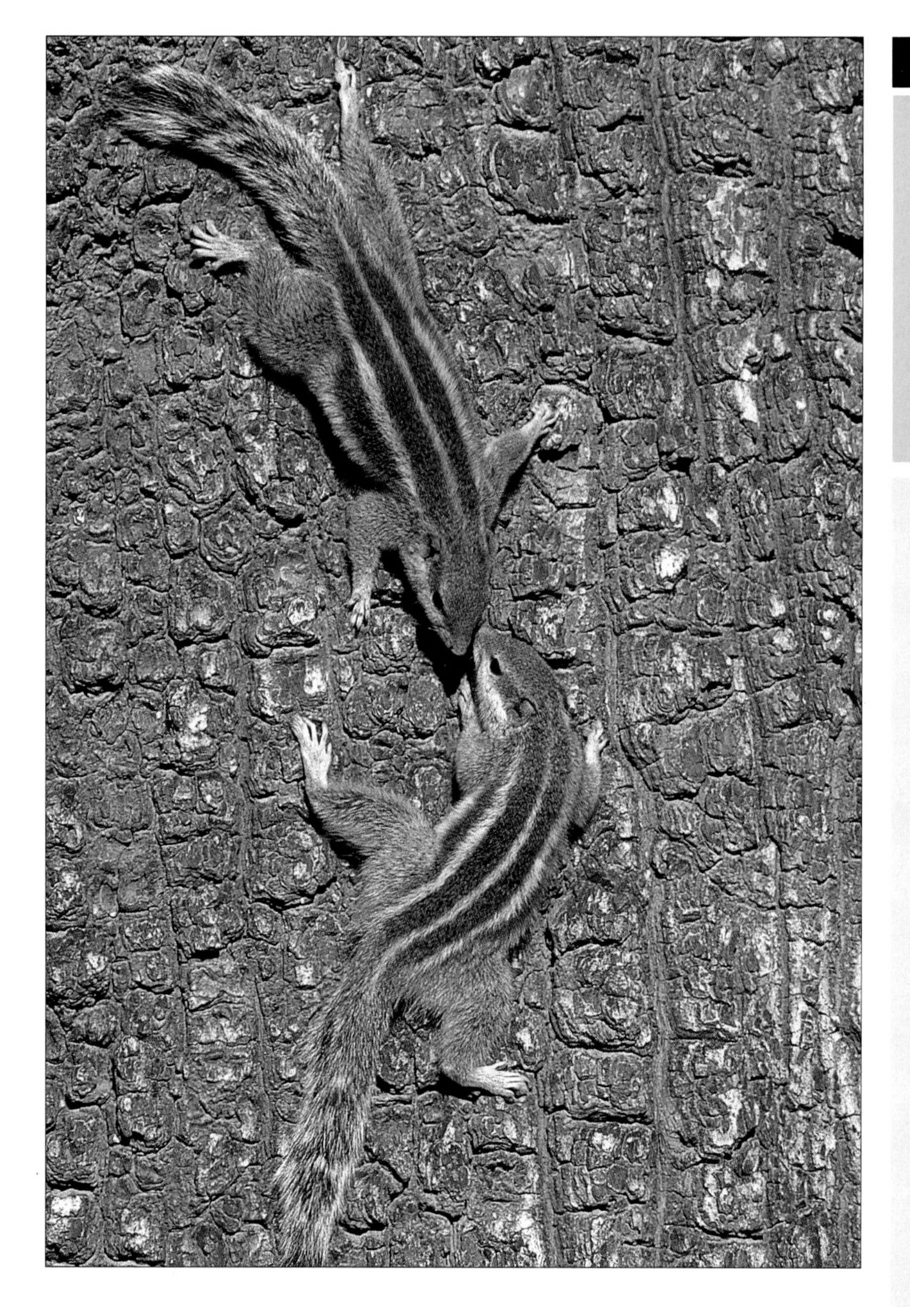

Five-striped palm squirrels, **Funambulus pannanti,** *northern India. Asian palm squirrels are found in a wide range of habitats, from palm forest to dense jungle.*

mates with her. Although mating is prolonged, up to 20 minutes, the male stays with the female for only a day. The female builds a nest on the branches of a tree or in a hollow. It is spherical and made of dry grass or other fibrous materials such as bits of cloth, animal hair and feathers.

The main breeding season is between March and September, but there may be 3 litters a year with one to five, usually three, in a litter. The young are born after 40–45 days and are naked, with pink skin. They measure about 4 inches (10 cm) long, of which more than half is tail. Their folded ears open at 7 days, and the eyes open at between 15 and 25 days. Weaning is at 25–30 days, and the young have grown a full coat of hair, shorter than in the adult but with the same color and pattern, by 5–8 weeks. They are then half the size of the parent. The young females are sexually mature at 6–11 months.

PALM SQUIRRELS

CLASS **Mammalia**

ORDER **Rodentia**

FAMILY **Sciuridae**

GENUS **Asian palm squirrels, *Funambulus*; African palm squirrels, *Epixerus*; oil-palm squirrels, *Protoxerus*; Berdmore's palm squirrel, *Menetes berdmorei***

SPECIES **10 species**

WEIGHT
⅓–2¼ lb. (0.15–1 kg)

LENGTH
Head and body: 4½–15¾ in. (11.5–40 cm); tail: 4½–14 in. (11.5–36 cm)

DISTINCTIVE FEATURES
Usually brown, olive, tawny or gray upperparts, often with pale stripes running down body; light brown, cream or gray underparts; paler or reddish brown head

DIET
Nectar, pollen, fruits, seeds and seed pods (mainly palm pods in some species); also insects and carrion

BREEDING
Age at first breeding: 6–11 months; breeding season: all year; number of young: 1 to 5; gestation period: 40–45 days; breeding interval: up to 3 litters per year

LIFE SPAN
Asian palm squirrels: up to 2 years; oil-palm squirrels: up to 6 years

HABITAT
Open palm forest and scrub to dense jungle; some species in and around human habitation

DISTRIBUTION
West to east-central Africa; Indian subcontinent; Myanmar (Burma) to Vietnam

STATUS
Most species common

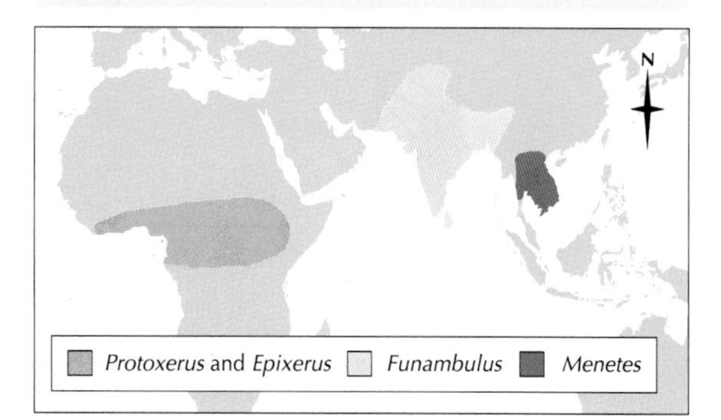

PALOLO WORM

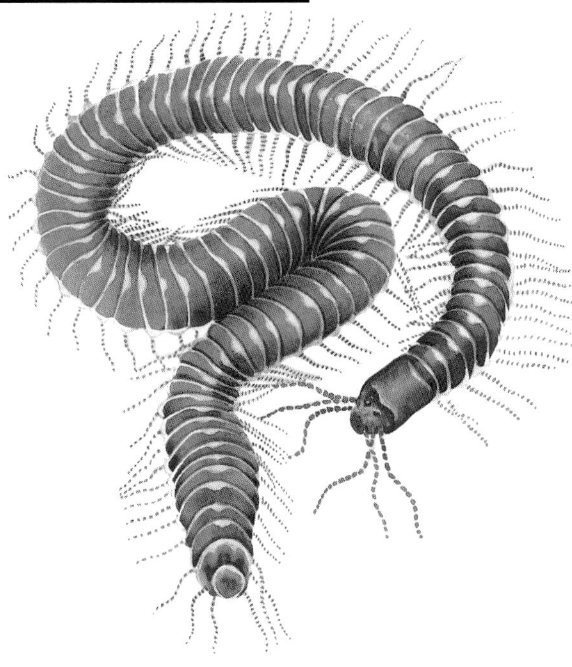

The spawning of palolo worms, during the last quarter of the moon in October and November, is eagerly awaited by local people. The worms are highly esteemed raw or cooked, and provide a valuable source of protein.

PALOLO WORMS

PHYLUM **Annelida**

CLASS **Polychaeta**

ORDER **Eunicida**

FAMILY **Eunicidae and Lumbrineridae**

GENUS AND SPECIES **Many**

LENGTH
Average 1⅓ ft. (40 cm)

DISTINCTIVE FEATURES
Very large size; ragwormlike body; many obvious segments, each with 1 pair of paddlelike appendages; appendages have gills

DIET
Assumed to prey on passing animals

BREEDING
Breeding season: last quarter of moon in October–November; mass release of epitokes (rear portion of palolo worms) containing sperm and eggs

LIFE SPAN
Not known

HABITAT
Coral reefs

DISTRIBUTION
South Pacific: Samoa and Fiji; Indonesia: Moluccas Archipelago

STATUS
Superabundant

TWICE A YEAR, AND WITH incredible regularity, the rear portion of palolo worms develop almost into another animal, and swarms of these invade the sea to breed. Palolo worms measure about 1⅓ feet (40 cm) in length, and, like their relatives the ragworms, are divided into a large number of segments, each with a pair of paddlelike appendages that bear gills. The head has several sensory tentacles. Males are reddish brown and females are bluish green.

Palolo worms live mainly on coral reefs off Samoa and Fiji in the South Pacific, and are caught by local people in vast numbers when spawning. They are considered a great delicacy.

Worms split during spawning

Palolo worms riddle coral, digging tubes in the reefs, in crevices or under rocks. They are very difficult to extract because their long, fragile bodies are firmly anchored inside these tubes.

Toward the breeding season the rear half of the body of a palolo worm alters drastically. The muscles and other internal organs degenerate while the reproductive organs in each segment grow. The limbs become more paddlelike and in due course the palolo worm backs up its tunnel, with the head innermost, until the modified part of its body protrudes. This part, known as the epitoke, breaks free and swims to the surface, complete with rudimentary eyes with which to navigate. The remaining front half stays in the tube and regenerates the lost portion.

These free-swimming portions of the palolo worms are actually little more than bags of eggs and sperm. On reaching the surface, these are discharged and the empty skins sink to the bottom to be devoured by fish. The eggs are fertilized as they float at the surface, and from them free-swimming larvae develop.

Even more interesting is that the epitokes are released at exactly the same time each year. When this happens, the sea becomes a writhing mass of millions of worms and is milky with eggs and sperm. Swarming is limited to the last quarter of the moon in October and November. Although occurring at different times of day on different islands, the peak of swarming is reached in less than an hour after the first worms emerge in any one place.

If the mechanism is obscure, the advantages are obvious. By releasing all eggs and sperm in one or two highly synchronized batches, the chances of fertilization are very much increased. Several similar species of worms, including palolo worms, have been reported spawning at the same time and place in the Moluccas Islands, Indonesia.

PANGOLIN

Pangolins are unmistakable animals, quite strange in appearance, the hair on their backs having been converted into large, overlapping, olive-brown scales covering the head, back, tail and legs. The underside of the body is, however, soft and hairy.

A pangolin's body is long and it has a long tail. The skull is also long and cone-shaped, and the snout is pointed, with a small mouth at the end and toothless jaws. The pangolin's long tongue can be thrust out to nearly 1 foot (30 cm) in some species. It has small eyes and hidden ears and its legs are short; the five toes on each foot have stout claws used in digging. In Africa there are four species of pangolins, or scaly anteaters as they are sometimes called, and in southern Asia there are three.

The giant pangolin, *Manis gigantea*, of equatorial Africa might grow to up to 6 feet (1.8 m) long, while the Cape pangolin, *M. temminckii*, reaches nearly 4 feet (1.2 m) in total length. Other African species are the long-tailed tree pangolin, *M. tetradactyla*, and the small-scaled tree pangolin, *M. tricuspis*, both of which range from West Africa to Uganda. These two species are around 3 feet (90 cm) long, including the tail.

The largest Asiatic species is the Indian pangolin, *M. crassicaudata*, which is 3½ feet (1 m) long. The Chinese pangolin, *M. pentactyla*, of Nepal, southern China, Hainan and Taiwan, and the Malayan pangolin, *M. javanica*, of Myanmar (Burma), Peninsular Malaysia and Sulawesi, are both under 3 feet (90 cm).

Ground dwellers and tree climbers

Most of these unusual, scaly beasts are arboreal and climb trees using their sharp claws and their tail, either wrapping the tail around a branch and sometimes hanging by it or using it as a support by pressing it against the trunk. The giant and Indian pangolins are both terrestrial, however, and live on the ground, although the Indian species sometimes climbs trees for safety when chased. All pangolins are active mainly at night, the terrestrial forms resting by day in burrows dug by other animals, the tree-dwellers resting in cavities in tree trunks. When on the ground, pangolins walk on the sides of their forefeet or on their knuckles, with their long claws turned inward. They often walk on their hind legs with the body raised semierect, the tail raised above the ground as a counterpoise.

Raiding ant and termite nests

This attitude, with the tail supporting the erect or semierect body, is also used when a pangolin is tearing open a termite nest with its long front claws, exploring the galleries of the nest with its long tongue. Its tongue is sticky and is flicked in and out. Ants are also eaten, including adults, pupae, larvae and eggs. The tough skin of the head protects the pangolin from attacks by soldier termites and from the stings of ants. The nostrils and ear openings can be closed, and the eyes are protected by thick lids. Ants crawling onto the body are shaken off, and those swallowed are soon ground by the thick muscular walls of the stomach and by the small pebbles that the pangolin swallows.

Babies ride piggyback

Very little is known about the breeding habits of pangolins because they often fail to breed in captivity. Mating takes place in late summer or the fall, but some species breed throughout the year. Females in the wild give birth to one young, rarely two, born after a gestation period of about 140 days. The scales do not harden until the second day after birth. Later the baby rides on the mother, clinging to her tail.

A Cape pangolin or scaly anteater, shown here in the Kalahari Gemsbok National Park in South Africa. Its heavy armor of overlapping scales protects it from the attacks of its termite and ant prey.

PANGOLINS

CLASS **Mammalia**

ORDER **Pholidota**

FAMILY **Manidae**

GENUS ***Manis***

SPECIES **7, including Indian pangolin, *M. crassicaudata*; Malayan pangolin, *M. javanica*; long-tailed tree pangolin, *M. tetradactyla*; and Cape pangolin, *M. temminckii***

ALTERNATIVE NAME
Scaly anteater

WEIGHT
5½–73 lb. (2.5–33 kg)

LENGTH
Head and body: 1–3 ft. (30–90 cm); tail: 1–3 ft. (30–90 cm)

DISTINCTIVE FEATURES
Unmistakable long-bodied animal with olive-brown scales across most of head, body, legs and tail; long, cone-shaped skull; no teeth; very long tongue; small eyes

DIET
Adults, pupae, larvae and eggs of ants and termites; probably some other insects

BREEDING
Age at first breeding: 1–2 years; breeding season: late summer to fall, or all year in some species; number of young: usually 1; gestation period: about 140 days; breeding interval: probably 1 year

LIFE SPAN
Up to 20 years in captivity

HABITAT
Forest, bush and savanna

DISTRIBUTION
Africa and southern Asia

STATUS
All species either endangered or at low risk

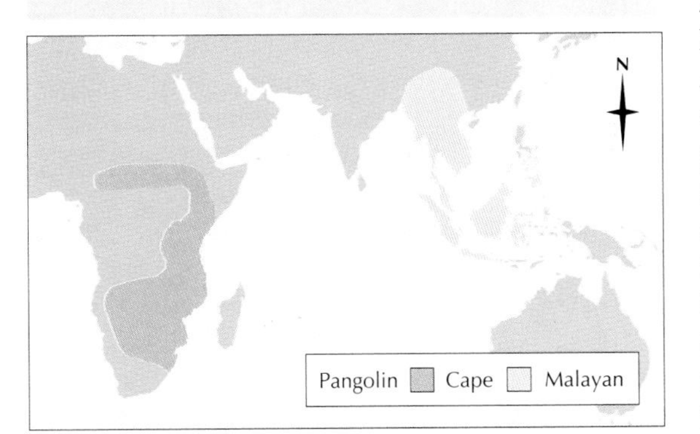

The Malayan pangolin is mainly arboreal and is able to hang from branches by its long tail. Here one is shown curled into a ball in a defensive posture.

Roll into a ball

The main threat to pangolins is probably humans, and all species are now classified as being either endangered or at low risk. Animals, such as large cats, sometimes examine them but are, it seems, put off by their scales. Pangolins are, however, killed by local people for their flesh, and their scales are used for ornaments and charms as well as for their supposed medicinal value. In Africa boys are sent into the burrow to put a rope around a pangolin's tail to drag it out. Its defense is to roll up and even a light touch on a pangolin's body makes it snap its sharp-edged scales flat. Some pangolin species, possibly all of them, can give off an obnoxious fluid from glands under the tail.

An animal puzzle

Pangolin skins brought back from Africa and Asia were known to the Romans and to the scientists of the 16th century and later. All were puzzled by them, as were the peoples in whose countries they lived. Arabs called the pangolin abu-khirfa, "father of cattle"; the Indians named it bajurkit, "jungle fish"; the Chinese name was lungli, "dragon carp"; and the Romans called it the earth-crocodile. The name pangolin is from the Malay peng-goling, the roller, derived from its habit of rolling into a ball.

PAPER WASP

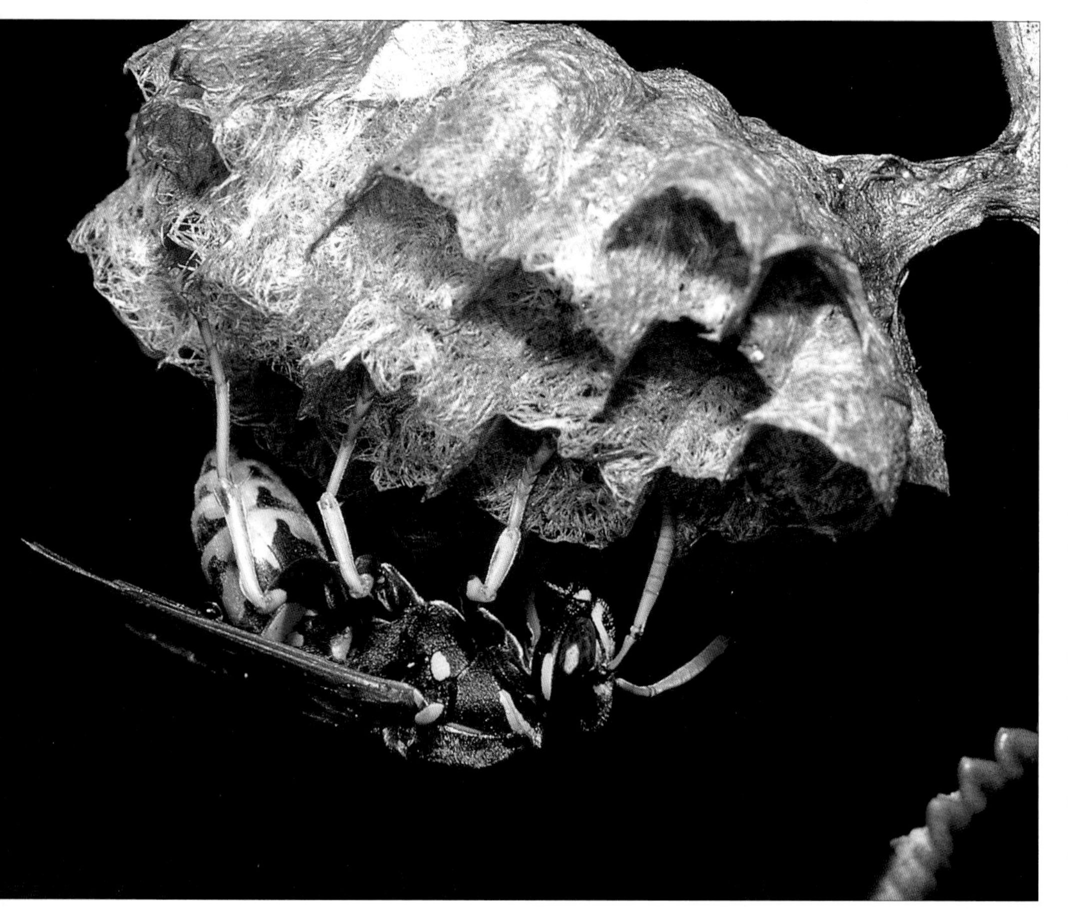

Paper wasps are so called because the queen constructs a small, umbrella-shaped nest of paper. The cells are open, rather than being covered by a papery sheath as they are with common wasps and hornets.

Paper wasp is the name given to wasps of the subfamily *Polistinae*. They are closely related to the common wasps and hornets of the genera *Vespula* and *Vespa* and there is little distinctive about their appearance. All are small, perhaps ⅖–⅘ inch (1–2 cm) long. The body is black with narrow yellow bands, and the antennae and wings are orange. However, the thorax and abdomen of the paper wasps are pear-shaped where they join, more like an hourglass than the rounded thorax and abdomen of the other wasps. Nor is the sting of a paper wasp, painful though it is, as toxic to humans as the sting of a common wasp or hornet. In habits paper wasps also differ in that they make small nests that are not enclosed by a papery sheath, and the cells containing eggs and larvae are open.

Paper wasps are found all over the world, in both tropical and temperate regions. They are particularly common in North America.

Umbrella-shaped nests

The life of the paper wasp is very easy to study because the nest is built above the ground, it is open so that all the activities can be watched and the colony is comparatively small. Some nests may number up to 500 wasps but there are usually less than 100. In small nests it is quite easy to mark all the wasps and to watch and record their behavior.

In tropical regions the nests survive year-round, but in temperate climates the workers and male wasps die in the autumn. Only the queens survive the winter, lying dormant in a shelter.

The following spring the queens emerge and seek a suitable place for their nests. As the nests are open, the site must be sheltered and not exposed to strong winds or strong sunlight. A common place for nests is under the eaves of houses and sheds or on branches. As these places often get warm, the nests are kept cool by the wasps fanning their wings, and they may bring water to regurgitate over the comb.

The queen constructs the nest from paper, which she makes by rasping wood from a tree with her jaws, chewing it and mixing it with saliva to make a paste. The first stage in building is to find a flat foundation on the underside of a roof or branch. The queen then creates a short stalk that holds the main part of the nest, the brood cells that make up the spherical comb, to the overhang.

Senior queen dominates

Sometimes in warmer regions several queens combine to build a nest and lay eggs, but eventually one queen asserts her authority and the others stop laying. If, however, the senior queen dies, one of the others will take her place. The ovaries of the lesser queens are small in comparison with those of the senior queen, and any eggs the lesser queens lay are destroyed. The senior queen is also looked after more carefully by the workers. Exactly how she manages to dominate the nest in this way is not known, but it may be by means of pheromones, chemicals given off from her body that affect the behavior and body functions of the other paper wasps.

The queen mates before her winter retreat and the sperm is stored until she lays her eggs the following spring. The workers develop from fertilized eggs but are infertile themselves because they are reared on a reduced diet. It is only later in the season that new queens are

PAPER WASPS

PHYLUM **Arthropoda**

CLASS **Insecta**

ORDER **Hymenoptera**

FAMILY **Vespidae**

SUBFAMILY **Polistinae**

GENUS AND SPECIES ***Polistes annularis*; others**

LENGTH
Adult: ⅖–⅘ in. (1–2 cm)

DISTINCTIVE FEATURES
Adult: black body with narrow yellow bands; orange antennae and wings; thorax and abdomen pear-shaped rather than rounded as in common *Vespula* wasps

DIET
Adult: flower nectar and fruit. Larva: pulped flesh of other insects; scraps of carrion.

BREEDING
Breeding season: autumn; number of eggs: rarely more than 100 larval cells in a nest

LIFE SPAN
From a few weeks to several years

HABITAT
Very varied; often nests on buildings

DISTRIBUTION
Worldwide

STATUS
Common

Nest of paper wasps, beneath an overhanging rock, northern Australia. In tropical regions such as this, the nests of paper wasps survive all year.

raised by feeding some of the larvae on a rich diet. At the same time males are reared from unfertilized eggs.

The first batch of larvae of the season are reared by the queen. However, once these first worker wasps emerge from their cocoons they will be left to tend the next generation. The larvae are fed on animal matter made up of carrion, caterpillars and other insects that have been chewed up by the adults. The larvae grow fat on this diet. After a few weeks the entrances of their cells are sealed with a thin layer of paper and pupation takes place. After emerging from their cells, the new workers take a meal from other workers and then start their life's work of raising new generations. The nest gradually increases as new cells are added, but in autumn larvae start to die, probably because of a failing food supply, and the colony dies off.

Before this happens the males and queens will have set out on their mating flights. The colonies cannot survive in temperate regions because wasps do not store food as honeybees do. In the Tropics and near-Tropics, however, the workers can become torpid during a brief period of bad weather and then become active again.

Cuckoo wasps

A few species of paper wasps are called cuckoo wasps and have a quite different life cycle. The details of their habits are not fully known, but it appears that the queens emerge from hibernation later than occurs in the other species. Each queen then searches for a flourishing nest and lays her own eggs in it. The reigning queen of the nest and the lesser queens, as well as the queen larvae and pupae, are killed. The cuckoo wasp eggs develop into males and females only, never workers, and these are raised by the host workers to maturity, when they fly off to mate.

PAPUAN TURTLE

Papuan turtles may actually be fairly common in Papua New Guinea and northern Australia, but are rarely seen, spending most of their time near the bottom of lakes and slow-moving rivers and streams.

THIS FRESHWATER TURTLE IS unique in that it is the only surviving member of its family and has paddlelike forelimbs like the sea turtles. It gets its name because for nearly one hundred years it was thought to occur only in Papua New Guinea. It has now also been found at a number of locations in the Northern Territory of Australia. In Australia it is called the pitted-shelled turtle. The Papuan turtle's anatomy is interesting because it is a link between the soft-shelled turtles of the family Trionychidae and the other cryptodiran turtles, that is those that bend their necks vertically when they withdraw their head into the shell. Fossil members of this family have been found in North America, Asia and Europe.

The maximum recorded length for the carapace (shell) of a Papuan turtle is 28 inches (70 cm), but most individuals are much smaller than this. It has a pitted appearance, hence the name by which this species is known in Australia. The shell does not have scutes (plates), as in the soft-shelled turtles, and is brown in color. The turtle has a prominent, proboscis-like snout and is also known as the pig- or hog-nosed turtle. The fore limbs are flatted as paddles to aid swimming, and end in two claws. The hind limbs have webbed toes, also an aid for efficient swimming.

Hunt near the bottom

Most of the life of Papuan turtles is spent in the water of slow-flowing rivers and streams, as well as in lakes, lagoons and occasionally estuaries, especially of the Fly River in Papua New Guinea. The flippers and the light, streamlined body suggest that they are active swimmers living in open water. However, these turtles are more often found in shallow water near the bottom, and one of the functions of the strong, broad claws is to anchor the turtles to rocks or logs.

The claws are also used to hold prey that cannot be swallowed whole. While the food is held firm, pieces can be bitten off by the jaws. Papuan turtles are omnivorous feeders. They eat the fruits of pandanus trees and figs that drop into the water, leaves, roots and the seeds of aquatic plants. In terms of animal matter, they are thought to concentrate mainly on slow-moving prey, particularly snails, insects and small fish. They also eat carrion.

Do not hatch until the rains

It is not known how many Papuan turtles exist in the wild. Turtles that live in lakes and rivers are often thought to be less common than they actually are simply because they are rarely seen except when the females emerge on land to breed.

PAPUAN TURTLE

CLASS **Reptilia**

ORDER **Testudines**

FAMILY **Carettochelydidae**

GENUS AND SPECIES ***Carettochelys insculpta***

ALTERNATIVE NAMES
Pitted-shelled turtle; pig-nosed turtle; hog-nosed turtle; Fly River turtle

LENGTH
Up to 28 in. (70 cm)

DISTINCTIVE FEATURES
Like soft-shelled turtles, lacks scutes (plates) on carapace (shell); large, proboscis-like snout; paddle-like foreflippers with 2 claws at end; webbed toes on hind limbs; brown carapace, which appears pitted

DIET
Fruits of pandanus tree, figs, leaves, roots and seeds of aquatic plants; also snails, insects, small fish and carrion

BREEDING
Age at first breeding: not known; breeding season: end of dry season (October to November in Papua New Guinea); number of eggs: about 15; hatching period: may not hatch until first rains; breeding interval: 1 year

LIFE SPAN
Not known

HABITAT
Slow-flowing rivers and streams; also lakes, lagoons and occasionally estuaries

DISTRIBUTION
Southern Papua New Guinea; northern Australia: Daly, Victoria and Alligator River systems, Northern Territory

STATUS
Not known

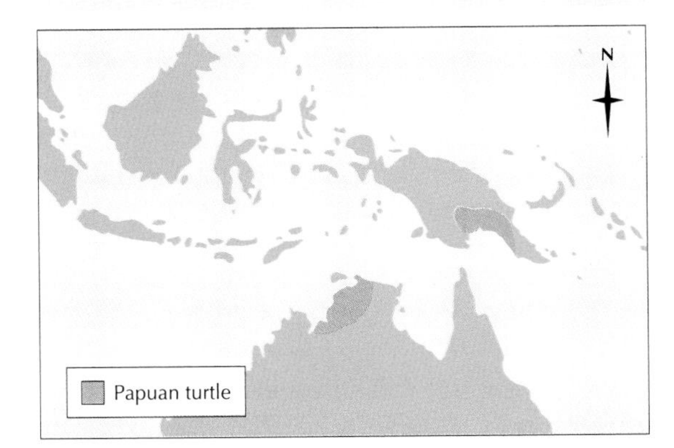

Papuan turtle hatchlings. Breeding in this species is unusual in Australia, the fully developed embryos waiting until the first rains before hatching.

The habits of these turtles in Papua New Guinea are not well known, although the emergence of females onto mud and sandbanks at the end of the dry season (October–November) to lay their eggs has been recorded. The turtles dig small holes or burrows in the mud or sand at the edge of a river in which they deposit about 15 eggs. Breeding in Australia is slightly different. Laying also takes place at the end of the dry season, but once the embryos have fully developed they remain quiescent, or inactive, within the egg and do not hatch until the first rains of the following wet season.

Underwater breathing

At intervals a Papuan turtle leaves the river bottom and swims steeply to the water's surface to breathe. However, like other aquatic tortoises and turtles, it gets a fair proportion of its oxygen from the water. It is thought that it does this partly through its skin because blood vessels can be seen near the surface of the soft skin. Moreover, when lifted from the water, a Papuan turtle expels water from its cloaca (a chamber into which the urinary, intestinal and generative canals feed), so exchange of oxygen and other gases could take place in the hind part of the intestine, as happens in some fish. It also seems likely that the turtle's mouth is a site for gas exchange, just as some fish and frogs use the lining of the mouth as a sort of lung. While the turtle is resting on the bottom, the throat can be seen pumping about 24 times each minute. Water is drawn in through a gap between the jaws and is ejected through the nose and mouth. It is possible that this water current may also enhance the sense of smell, for chemicals will be drawn through the nose.

PARADISE FISH

The paradise fish's long fins may act as organs of touch in the dark, muddy surroundings of its native habitat.

GOLDFISH HAVE BEEN KEPT as household pets for a long time, but it was in the mid-19th century that home aquariums first became popular. Initially only marine animals and a few freshwater fish were kept in them. Then in 1869 the paradise fish, *Macropodus opercularis*, was brought to Paris from its native Southeast Asia. Before long it had reached England and other European countries, and was subsequently taken to the United States. At first aquarists were concerned that the paradise fish would injure their goldfish. However, the paradise fish proved a popular species and started the fashion of keeping "tropicals."

The paradise fish can grow up to 3½ inches (9 cm) long and has a body that is flattened from side to side. It has flowing dorsal and anal fins, a large, rounded tail, small pectoral fins and pelvic fins about the same size as the pectorals and lying beneath them. The male paradise fish is brown to greenish gray with marbling on the head and a large blackish spot ringed with orange on each gill cover. The flanks are banded bluish green and carmine. The fins are reddish, the pelvics are white-tipped and the dorsal and anal fins have dark spots. The female is similar in color to the male, but paler.

The paradise fish ranges from Korea through eastern China, including Taiwan, to southern Vietnam. The spike-tailed paradise fish, *Pseudosphromenus dayi*, similar but slightly smaller, has a similar range but does not go so far south. A third species, the round-tailed paradise fish, *M. chinensis*, also small, with two longitudinal bands on the flanks, ranges from India and Sri Lanka through Myanmar (Burma) to southern Vietnam.

There has been much scientific debate about two points concerning the paradise fish: whether it is the wild form or one bred in China, and what is the purpose of its bubble nest. On the whole, it seems that the fish kept in aquariums is much the same as the fish that is wild in Chinese rice fields. The aquarium breeds have been only slightly altered from the wild forms, though there are also special breeds. There is, for example, a dark variety, the black paradise fish, *M. concolor*, and an albino strain that is white with pink eyes and pink bands on the flanks, which breeds with true paradise fish.

Bubble nesters

Along with four other families of the order Perciformes, including climbing perches, gouramis and fighting fish, paradise fish are known as labyrinth fish. All these fish have a small respiratory organ known as a labyrinth organ located above each gill chamber. This enables the fish to extract minute quantities of oxygen from air bubbles that they take in through the mouth and allows them to live in water that contains virtually no oxygen.

As in many other labyrinth fish, paradise fish build nests of bubbles. The male blows out bubbles of air and mucus, which rise to the surface where they form a raft. The colors of the male become brighter as the breeding period draws near, and the female becomes paler. When the female is about to spawn, the male wraps himself around her while she lies just under the bubble raft. As the eggs are being laid, the pair make a barrel roll, which brings the female upside down with the male still wrapped around her, and he fertilizes the eggs as they leave her body. Then the male releases the female and trembles for a few seconds before gathering any slowly sinking eggs in his mouth, rising to the underside of the bubble nest and spitting the eggs onto it.

PARADISE FISH

CLASS **Osteichthyes**

ORDER **Perciformes**

FAMILY **Belontiidae**

GENUS **12 genera**

SPECIES **46 species, including *Macropodus opercularis* (detailed below)**

LENGTH
Up to 3½ in. (9 cm)

DISTINCTIVE FEATURES
Long, flowing dorsal and anal fins; large, rounded tail; ground color brownish or greenish gray; upper surface of head and nape marbled brown, black and olive green; large blackish blotch on each gill cover; carmine red and blue-green bands on flanks; bright red fins; female paler than male

DIET
Small aquatic animals, including small fish

BREEDING
Number of eggs: up to 500; hatching period: often within 24 hrs

LIFE SPAN
Not known

HABITAT
Streams, paddy fields and ditches; can colonize stagnant water low in oxygen

DISTRIBUTION
Southeast Asia north to Korea and Ryukyu Islands (south of Japan)

STATUS
Not threatened

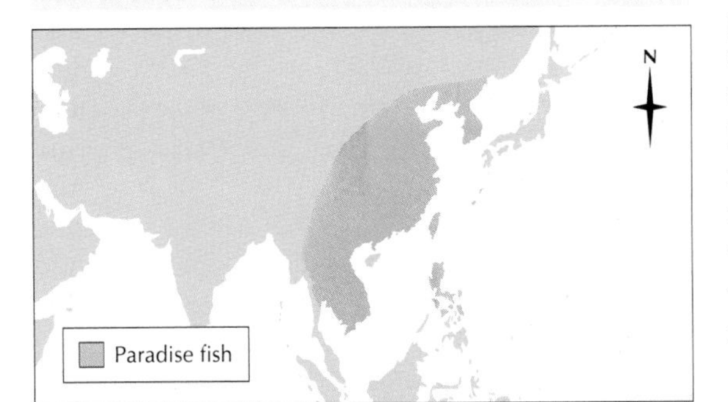

Unlike other bubble nesters, which have eggs that are heavier than water, those of the paradise fish mostly float upward when laid. Only a small number fail to do this and slowly sink. When the clutch is complete, the male blows more bubbles to make a second layer under the eggs, sealing them in. This process may be repeated.

The first tropical fish to be brought to Europe, the paradise fish is an aggressive and hardy fish that can survive in water that has virtually no oxygen content.

Scientists differ as to the purpose of the bubble raft. One theory is that it protects the eggs from the heat of strong sunlight beating down on the rice fields. Another is that it shades the eggs from strong light. A third is that it protects the eggs from bacteria. The fourth suggestion, and the most likely, is that it keeps the eggs together, and also the fry when they hatch, making it easier for the male to guard them. The eggs hatch in 2 days at a temperature of 80° F (26° C).

Paradise fish can respond to sounds of frequencies between 2,640 and 4,700 cycles per second. It is possible that they make sounds, inaudible to human ears, that stimulate breeding.

Aggressive tendencies

Little is known about the predators of paradise fish in the wild. More is known about their pugnacious nature in aquariums, where they tear the fins of other kinds of fish if placed in a mixed tank. Paradise fish feed on any small animals, being very predatory, and they readily attack members of their own kind. Aquarium owners must take care to keep male and female apart until they are ready to breed, and it may happen that, after mating, the female is savaged by the male if she has too little space to escape him.

The reason for such behavior is not immediately clear. Perhaps clues can be found from experimental work carried out on paradise fish in the laboratory. The fish eat more food when grouped together but grow more quickly when placed in tanks alone. Their aggression may be purely intended to keep each fish spaced out, to give them all adequate growing space, or it may be a natural means of controlling numbers.

Index

Page numbers in *italics* refer to picture captions.
Index entries in **bold** refer to guidepost or biome and habitat articles.

Page numbers in *italics* refer to picture captions. Index entries in **bold** refer to guidepost or biome and habitat articles.

Page numbers in *italics* refer to picture captions. Index entries in **bold** refer to guidepost or biome and habitat articles.